Applying the Scientific Method to Learn from Mistakes and Approach Truth

Applying the Scientific Method to Learn from Mistakes and Approach Truth

Finlay MacRitchie

CRC Press
Taylor & Francis Group
Boca Raton London New York

CRC Press is an imprint of the
Taylor & Francis Group, an **informa** business

First edition published 2022
by CRC Press
6000 Broken Sound Parkway NW, Suite 300, Boca Raton, FL 33487-2742

and by CRC Press
2 Park Square, Milton Park, Abingdon, Oxon, OX14 4RN

CRC Press is an imprint of Taylor & Francis Group, LLC

Library of Congress Cataloging-in-Publication Data
Names: MacRitchie, Finlay, author.
Title: Applying the scientific method to learn from mistakes and
approach truth / Finlay MacRitchie.
Description: First edition. | Boca Raton: CRC Press, 2022. |
Includes bibliographical references and index.
Identifiers: LCCN 2021038876 (print) | LCCN 2021038877 (ebook) |
ISBN 9781032183367 (hardback) | ISBN 9781032183305 (paperback) |
ISBN 9781003254065 (ebook)
Subjects: LCSH: Science–Methodology. |
Resilience (Personality trait)–Australia. | Education–Social aspects–Australia. |
Australia–Social conditions. | Australia–Politics and government–1945–
Classification: LCC Q175 .M2349 2022 (print) |
LCC Q175(ebook) | DDC 001.4/2–dc23/eng/20211021
LC record available at https://lccn.loc.gov/2021038876
LC ebook record available at https://lccn.loc.gov/2021038877

ISBN: 978-1-03-218336-7 (hbk)
ISBN: 978-1-03-218330-5 (pbk)
ISBN: 978-1-00-325406-5 (ebk)

DOI: 10.1201/9781003254065

Typeset in Palatino LT Std
by Newgen Publishing UK

Contents

Preface

The title of the book is intended to incorporate two concepts. One is that learning from our mistakes is a simple representation of the scientific method. The other is that although the aim of science is to search for truth, it can never claim to attain it although it can claim to approach it. Many people believe that they know the truth. Some may have formed their beliefs as a result of seeking information and thinking deeply about issues. Others may have been convinced by persuasive arguments. There are some who are easily influenced and uncritically believe what they are told; in other words, they are brainwashed. Is it better to steadfastly believe that we know the truth or, as in the scientific method, to always retain doubts and never assume we hold the absolute truth? Many would side with the former. Isn't it better to feel sure and to have complete confidence in your belief than to never assume that you know the truth? What about if the belief you hold is false? Just because you hold a strong belief doesn't mean that it is necessarily true. What is the criterion for believing that you hold the truth? There may not be one. On the other hand, scientists, although never claiming to know the absolute truth, do have a criterion for establishing if they are moving towards it.

The hypothetico-deductive scientific method creates a hypothesis which is then severely tested. If the test refutes the hypothesis, it is rejected and a search for a new one is commenced. The new hypothesis may make use of what was learned from the failed one. If the test does not refute it, it is considered to be corroborated. Although it can never be assumed that the absolute truth has been attained, more confidence is justified that we are moving towards it. Thus, we have a criterion for getting closer to the truth.

There are people who believe that they possess the truth in regard to a certain way of thinking or to an ideology. If the belief turns out to be false, then the gift of life that has been bestowed on them may have been wasted. It would have been better to have gone through life searching for truth rather than basing their actions on a false belief.

In this book, the scientific method is applied to analyse some current issues. Two important ones are given special consideration. One is anthropogenic global warming and the other is enforced lockdowns due to the coronavirus. Two hypotheses have resulted. One is that there is a significant contribution of human activities to warming of the planet. The other is that the best approach to dealing with the virus is to impose community lockdowns. Considerable disruption and economic destruction have resulted from decisions based on the two hypotheses. These detrimental effects have not resulted from the problems themselves but from the policies that have been introduced to supposedly counter them. The policies have not been based on science although this is what is often claimed. Both hypotheses have been refuted but the policies have been dogmatically preserved.

How does a society guard against its members adopting false beliefs? It must begin with the education system. Students need to have instruction on how to think critically and to understand how the scientific method works. Instead of adopting this approach which could result in a society whose members are free thinkers, we are seeing the imposition of ideology. Our history should teach us about the destructive effects of totalitarian ideologies. Ideologies from the extreme left (communism) and the extreme right (Nazism) have both caused death and destruction in the last century. Can we learn from this or do we really want to?

There is presently an ideology that has been named "wokeism" that is pervading society. There are no concentration camps or gulags yet but there is a movement in that direction. This is evidenced by intolerance of opposing viewpoints and what has been called the "cancel culture" that has infiltrated society. The flames of this ideology have been fanned by the education system and have spread into the media, the bureaucracy, the judiciary, and large portions of the general public. The media which we rely on to get a balanced reporting of current events has become heavily biased towards the political left. The judiciary which we expect to impartially apply the law are increasingly legislating from the bench. Some of our political representatives, supposedly our servants and charged with defending the interest of the public, often appear more concerned with having their snouts in the trough and being careful not to invite trouble by offending anyone. Some of our bureaucrats seem to be more concerned with increasing their power and finding ways to frustrate peoples' needs rather than trying to help them, which should be their job.

All of us are human and we all make mistakes. There are two crucial questions. The first is "Are we capable of recognizing our mistakes?" The second is "Do we learn from them so that we can correct them and make things better?" Some of the political decisions that have been made are deliberated on to reflect on whether they were good ones or mistakes.

Citizen military training (cadets, national service) in Australia has been phased out. This shows that we don't belong to a belligerent country and that we are virtuous.

Tertiary education has been made available to everyone, irrespective of merit. Surely this is fairer? Could there be any negative consequences?

Marriage laws were changed to allow people of the same sex to marry. That's fairer surely? Were the people who had made vows based on their understanding of traditional marriage consulted or was the decision made by everyone, including those who would not be affected and wouldn't give the matter deep thought?

We have protected our environment and put in place policies for phasing out industrial emissions to save the planet. That has to be good, right?

We are replacing those evil pollution-emitters by renewable energy sources such as wind and solar. What could go wrong?

These are some of the issues that will be discussed. To address them, we will try to use rationality and application of the scientific method.

About the Author

Finlay MacRitchie was a professor in the Department of Grain Science and Industry, Kansas State University from 1997 to 2009. He is presently Professor Emeritus in that department. Prior to this, he was a research scientist in the Commonwealth Scientific and Industrial Research Organization (CSIRO) of Australia. He has spent short periods of time as Visiting Professor at the University of Chile and the Federal University of Rio de Janeiro, Brazil and as Senior Research Fellow at the Agricultural University, Wageningen, The Netherlands, the University of Paris V, the University of Lund, Sweden, and the University of Tuscia, Italy.

Professor MacRitchie has published more than 150 papers in refereed journals and four textbooks – *Chemistry at Interfaces* (Academic Press, 1990), *Concepts in Cereal Chemistry* (Taylor & Francis, 2010), *Scientific Research as a Career* (Taylor & Francis, 2011), and *The Need for Critical Thinking and the Scientific Method* (Taylor & Francis, 2018). He is listed as an Institute for Scientific Information (ISI) highly cited researcher and is included in a list of the top 2% of researchers in the world by a recent Stanford University survey.

He has been a member of the editorial boards of Advances in Colloid and Interface Science, Cereal Chemistry, and Journal of Cereal Science and Editor-in-Chief of Journal of Cereal Science.

Professor MacRitchie's awards include the F.B. Guthrie Medal of the Cereal Division of the Royal Australian Chemical Institute (RACI) and the Thomas Burr Osborne Medal and George W. Scott Blair Memorial Award of the American Association of Cereal Chemists (now Cereals and Grains Association).

chapter one

The Importance of Building Resilience

Early Memories

Currently, there seems to be a public perception that young Australians of today and possibly those from other Western countries (especially some males, not all) are lacking in the resilience that was characteristic of earlier generations. The term "snowflakes" is often heard. This was not a term you could imagine being applied to the pioneers who contributed to building this successful and vibrant country. Nor could it be thought to describe those who volunteered to fight in World War I or who battled the enemy in Europe and the Pacific islands in World War II. What has changed? I was a five-year-old kid when World War II started so I will have a try at answering the question in terms of what has happened since that time and will base it on my own experience.

I grew up on a dairy farm in Northcliffe, a small town in the southwest of Western Australia. The population was only a few hundred and this may have even included the cats and dogs. My parents had migrated from Scotland several years before I was born. On arrival, they joined groups of other migrants from the UK to be thrown into what was then virgin country. These pioneers were expected to clear the big-timber forest and to establish farms, which they did. The government's group settlement scheme provided interest-only loans to enable them to purchase livestock and machinery in order to inaugurate their farms. My memories of the people of that time are of hard workers who cooperated willingly with each other and faced a daunting future with optimism. They met the obstacles that confronted them with a sense of humour and triumphed to form a successful community. An example of their self-effacing humour was that they always boasted that the town of Northcliffe was not on the map. This was true. The maps at the time showed the road and railway line from Perth going south and finishing at Pemberton, the nearest town some 30 kilometres to the north.

I recall a couple of amusing incidents when I was probably only about five or six years old. As is the custom today, children would celebrate their birthdays with parties and invite their friends. Since I lived in the country,

DOI: 10.1201/9781003254065-1

we didn't have next-door neighbours as such. Our nearest neighbours were at a distance of several miles. At one party I attended at a neighbour's home some six miles away, it was quite a hot summer's day. I saw that the children were going to a water tank to get drinks. One of the children went to this tank, filled a glass, and brought it to me. I thought the red drink that I was brought was the most delicious thing I had ever tasted and I thought "what a wonderful tank!" I wanted more of that so, a bit later, I said I was thirsty. One of the children went to the tank and filled a glass. Instead of the succulent red drink that had come out of the tank before, I was brought a glass of colourless, tasteless liquid. I was thirsty but not that thirsty. How could it be that a tank that had held this delicious red drink before had now changed to one that only produced this tasteless liquid? Of course, what I had witnessed and didn't understand at the time was that, before going to the tank, the kids had poured a small amount of raspberry cordial into the glass.

There was another story I recall that involved the same neighbours. We always had cats as pets. One of the cats was a female tabby and she would occasionally disappear. After being absent for several months, she would return and resume her life as if nothing had changed. We happened to relate this to our neighbours. "How strange" they said, "we have a cat that does the same. It goes away and then, about six months later, it reappears perfectly healthy". "What sort of cat is it?", we asked. "It's a female tabby", they replied.

When I was five years old, I went to a school that was only a couple of hundred yards from our home. On reaching the age of ten, this school closed because the number of students had declined below the critical value which was eight. The nearest school was then in the town of Northcliffe, some five miles away. Accompanied by my next older sister, I had to ride a bicycle to school along an unsealed gravel road, usually highly corrugated. Before starting the journey, I had to milk about four cows by hand and feed some calves before having breakfast and setting off for school. One thing that has always puzzled me was that on the way to school, it was mostly uphill and, on the way home, it was mostly uphill. I was never able to explain this topographical paradox. In winter, it could be bitterly cold in the mornings so that I could scarcely feel my hands on the handlebars. In summer, the afternoons could be extremely hot and, on occasions, we would have to fight our way through bush fires. Occasionally, we would see large snakes slither across the road. On one occasion, I saw the tail of a gigantic snake disappearing into the bush. It must have been a python. I didn't think there were pythons in that area but I learned later that there was at least one species.

Bush fires (called wildfires in the U.S.A.) were the only natural disaster that we had to contend with. Our house was built with timber and, as it was surrounded by trees and bush, was quite vulnerable. Bushfires were regular events in summer but usually we could cope. There was one occasion, however, when we were in great danger. The seriousness of a bush fire depends on the fuel load (trees and bush), high temperature, and low humidity but perhaps the most dangerous variable is strong wind. There was one occasion when it was very windy. We saw smoke in the distance but, suddenly in the space of a few minutes, the fire was upon us. With the strong wind, it seemed to have travelled through the tops of the trees. Our defence preparations were to have buckets of water drawn from a nearby well and hessian bags that had been soaked with water. Our house caught fire in several places but we were able to extinguish the flames using the hessian bags and dousing with buckets of water. The haymaking season had just finished and we had a shed full of hay quite close to the house. Embers landed on the hay and we were unable to prevent the fire from spreading. The hay did not go up in flames as it was tightly packed. However, it smouldered throughout the night and, next morning, all that remained was a heap of ash from the hay and the shed.

While on the subject of hay, I should mention a chore that helped me develop tolerance for discomfort. In haymaking, after the grass is cut, it is allowed to partially dry and then raked into windrows. Then a pitchfork is used to pile the grass into conical heaps which we called stooks. Later, these can be easily loaded onto a cart and transferred to the hay shed. With both hands needed on the pitchfork to make the stooks, we would be subjected to clouds of bush flies that tried to land on our faces. The only way to cope was to continually shake your head to keep them off. In Australia, some people used hats with corks dangling from them to deal with the problem but we didn't ever try that. It was a great exercise for developing endurance while maintaining our composure.

Nowadays, when I encounter ten-year-old boys, I find it difficult to imagine that they could do what I did at that time. On reflection, I suppose it was a fairly hard life. I didn't think that it was particularly hard. I just accepted that was the way it was. On the farm, we had no running water and no electricity. We got our potable water from tanks that collected water from rain falling on the roof. Occasionally, birds would foolishly build their nests in the guttering so that when there was a downpour, the poor things would be swept into the tanks, to be discovered some time later. Our lighting was lamps using kerosene or methylated spirits. Wood was used for the kitchen stove and the living room fire during winter. We had no air conditioning so, in the hot summer nights, we would leave doors open for cooling. Occasionally, bats would enter and fly around the living room, knocking against the walls and anything else in their paths. Wood was plentiful and one of my chores was to chop the firewood for the stove and the living room fireplace.

My reason for recounting these details of my early life is not to attract accolades. It wasn't such a hard life. We always had enough food, albeit very plain, and, as a member of a large family of which I was the youngest, I enjoyed interaction with my siblings. In the evenings, we read books, played cards and other games, listened to radio programmes, and played records on the gramophone. What I now realize and acknowledge with gratitude is that my experiences stimulated my mind and also helped me to develop resilience. This has served me later in life to face difficult situations, which everyone inevitably encounters. Many young children these days are not provided with the conditions required to build strong resilience. In addition, many of the activities in which they participate, such as using smartphones, are passive and not intellectually stimulating. Another memory I have is that, before starting school, one of my sisters gave me some lessons in reading. Looking back, I find it amazing that I picked up the skill with so little instruction. Soon, I was able to read simple texts. I know that I am not a genius so it makes me realize how easy it is for children to pick up knowledge when their minds are open.

At the age of 12, I was awarded what was called an Inspector's Scholarship. This was awarded to a restricted number of students who lived in regional areas that were not close to secondary schools. The scholarship made it possible for my parents to send me to the nearest secondary school, Bunbury High School. This was over a hundred miles away and necessitated boarding in Bunbury during school terms. My new life helped me to develop independence and self-reliance which again was part of building resilience.

I did fairly well in my studies but I should mention one subject that I took for one term and that was Latin. It is noteworthy because it was the only subject in my whole academic career that I scored nought out of a hundred in the final exam at the end of term. There was a rumour going around the school that Latin was essential if one decided to become a pharmacist. In fact, I didn't know of anyone in the school at that time who took up a career in pharmacy. In any case, I am sure that it was possible to become a pharmacist without knowing much Latin. It was one of those rumours that typically gets spread around and swallowed by everyone without it having any basis. I realize now that such shibboleths pervade society so that many widely held beliefs have strong influences on people but are based on falsehoods.

For our first class in Latin, the teacher (a lady) burst into the classroom and screamed, "humanum est errare". This got my attention. I thought this is different. I think I might enjoy a class with this eccentric woman. But I was to be disappointed. No sooner had she reached the desk than she began filling the blackboard with conjugations and declensions. I had never heard of these words and still don't know what a declension is. I had come from a one-teacher school out beyond the black stump, a bit to the left. The kids from the town of Bunbury were more at home and knew a lot more than me. I lost interest immediately and reverted to making paper aeroplanes and flying them around the room when the teacher wasn't looking. This was practically all the time as she was completely immersed in writing furiously on the blackboard.

At the end of term, we had an exam. I recall waiting outside the classroom before being ushered in to take the exam. There was one kid who approached me and seemed to be in a sorry state. He was almost in tears. He said to me "I don't know anything for this exam". In those days, I was a bit of a devil. I was not the nice, kind perfect human being that everyone knows me as today. I was a really nasty kid. So when this little fellow came to me with his tale of woe, I said, "Relax! There's nothing to worry about. There's only one thing you need to know for this exam. It's humanum est errare. If you know that, you'll sail through this exam". So, he wrote it down and I explained what it meant. Then, for the few minutes before going into the room, he was busily trying to memorize this stupid thing.

For the next class, the teacher came in with the exam marks. She read out the names of everyone and their marks. "MacRitchie, nought". "Stand up!" She made me stand up and continue standing for the duration of the class. A little later, she called out "Mills". He was the kid who I had "coached". I had taught him everything I knew – about Latin that is. He got nought as well and was made to stand up for the remainder of the class. While we were both standing, I happened to look across at my protégé. There were tears streaming down his cheeks. He felt terribly humiliated. I wasn't fazed at all. In fact, I derived a certain notoriety from it and I thought that the girls in the class might notice me as I felt that they hadn't noticed me up till then. That didn't work however. The girls in the class were mostly goody two-shoe types. They wouldn't dream of being associated with a kid who had got nought in an exam and been shamed by being made to stand for the whole class. For my protégé, it was different. I had done well in all the other exams. It was sheer bloody-mindedness that I didn't do well in this one. But my protégé just didn't have the capacity to learn and I think he was taken out of the school soon after.

As I recall this incident, I feel it was an important one in my development. I think it was one of the first times in my life up till then that I had felt compassion. I do remember an earlier incident when I killed a bird unintentionally. I was playing a silly game in a small orchard we had next to our house. The game consisted of gathering stones, whirling around and hurling them as you would when throwing a discus. On one throw, something didn't seem right as I sensed that the stone had hit an object. When I investigated, I found a dead bird at the foot of one of the fruit trees. After so many years, I still remember the feeling of sadness I experienced then.

I felt a similar sadness for my poor schoolmate. Compassion, I believe, is an attribute that is pretty much absent in young kids. It is something that we acquire during our journey towards adulthood. At least, most people do. Some don't but that is a topic that would need to be covered in another whole book.

I recall an incident when I must have been five or six years old. My father served in both World Wars I and II. In World War I, he had been in a Scottish regiment, the Seaforth Highlanders. We had a book that had the names and details of all who had served. I was fascinated by this book. My father's name was there and next to his name was a number. While the family was seated in the living room one evening, I innocently asked, "Is that the number of Germans my father killed?" What followed was pandemonium. Two of my sisters were escorting my mother to her bedroom. Someone was running to fetch a glass of water. I couldn't understand what was happening. "What's all this fuss about?", I was thinking.

I felt great empathy for my protégé. I thought it was most unfair to subject the poor kid to such humiliation. The teacher obviously failed to understand that he didn't have the intellectual capacity required so it was cruel to humiliate him in that manner. The lesson I learned and now realize was that the capacity for compassion is an essential quality for a balanced human to develop. I also don't wish to give the impression of denigrating Latin. I simply recount my experience at the time. Later, I was able to appreciate at least a little of the great contribution that Latin has made to Western culture.

My time at high school passed without much difficulty. One activity I should mention was that I participated in the School Cadets, which existed at that time. It included spending time on the rifle range for target practice. The rifle we used was a 0.303 Enfield, the standard rifle that had been used in World War II. In addition, camps of a week or so duration were held and included drill practice and long marches, sometimes at night. When I completed secondary school, I decided to take time off before starting university. The time off was planned to be two years but it turned out to be three because I was called up for National Service which was compulsory for 18-year olds at that time.

Pre-university Days

The three years enabled me to work and save enough money to cover my living expenses for my first year at university. I didn't want to ask for support from my parents as they were far from well-off. I was banking on getting a Commonwealth Scholarship at the end of my first year. In a sense, I burned my bridges and this gave me the incentive to succeed. I saved just sufficient money to cover my living expenses at the university hostel for my first year. Fortunately, there were no tuition fees at the time. The University of Western Australia was claimed to be a free university and so it was. The only expenses I incurred were for minor ones such as student union fees and, of course, textbooks and the like. I took various jobs during my pre-university three years. A land settlement scheme for returned servicemen (from World War II) was in operation at the time and there were ample opportunities for work in addition to helping on my parent's farm. One job involved using two horses to pull a heavy iron frame to level the ground after bulldozers had churned it up when clearing land by moving trees and logs into stacks prior to burning. Another source of work was caretaking of properties. Some of the returned servicemen who had been allocated farms were ill-equipped to manage them. Some had no experience of rural and farming life, were unable to cope and had to leave. My caretaking activities included milking cows by machine and looking after livestock, mainly pigs and calves until they could be transferred to other viable farms. Another source of work was on tobacco farms and this involved planting and harvesting. The tobacco industry was part of the returned servicemen's land settlement scheme. However, this industry soon failed and had to be abandoned.

In my third year after leaving school, I was called up to do National Service training which was compulsory at that time. We could choose which of the three services to enter and I opted for the Air Force. This meant six-month training at the RAAF station at Pearce in Western Australia. The first three months were devoted to what was called Ground Combat, essentially infantry training. The idea was that, in the event of an attack on the station, national servicemen would be part of the defence. We were given instruction in the use of weapons which at that time included the 0.303 Enfield rifle and machine guns, the Bren and Thompson (Tommy gun), grenades and mortars. Parade ground drill was an important part of our activities and other training involved deactivating booby traps and participating in mock battles. Our main activity was target practice using the rifle on the firing range.

There were many anecdotes I could relate about my National Service experiences but I will mention just a few. One exercise was to use hand grenades which were thrown from a trench. A hand grenade has a clamp with a pin to hold it in place. When the pin is removed, the grenade is kept safe by firmly holding the clamp. When the grenade is thrown, the clamp is released and there is a time of a few seconds before the grenade explodes. One of the national servicemen was obviously someone who had not played much sport and didn't have the coordination to throw (actually bowl with a straight arm). When his arm reached the highest point, the grenade simply dropped from his hand. Fortunately, it didn't roll back into the trench. The sergeant yelled, "Get down!" We all crouched in the trench, there was an explosion and we were showered with sand and debris.

On most mornings, there would be a parade in which all station personnel participated with rifles and bayonets. An order would be given "Squadron will fix bayonets! Fix!" On this command, the bayonet would be withdrawn from the scabbard and held in position near the muzzle of the rifle. The next command would be "Bayonets!" and the bayonet would be clamped to the rifle. One morning, the command was given to "Fix!" but then there was silence. We all wondered what was going on. The Warrant Officer in charge of the parade was very observant and noticed that one of the national servicemen had not withdrawn his bayonet. The previous night, some of his "friends" had filled his scabbard with chewing gum and the bayonet had become stuck. Two Non-Commissioned Officers (NCOs) came over and, after several minutes, managed to extricate the bayonet and the parade was able to continue.

One of the weapons we used was the PIAT (Projector Infantry Anti-Tank). This weapon was fired from the prone position unlike the Bazooka which was a bit lighter and could be fired from the standing position. For the firing practice, we were organized into two lines with the target being empty 44-gallon drums that were placed at a distance of about a hundred yards. When the first one in my line fired the weapon, we expected him to get up but he just lay there. The sergeant was shouting "AC Hartree, get up!" AC was our rank, aircraftsman, the lowest rank in the Air Force, equivalent to Private in the Army. It so happened that the Commanding Officer of the station was Air Commodore Healy. We would refer to him as AC Healy, one of us. So the sergeant was shouting "AC Hartree, get up!" but AC Hartree just lay there motionless, apparently dazed. They had to get two volunteers to help him to his feet and he staggered away. All of us who were lined up to fire this thing were thinking "This is going to be fun". It was not that it had such a kick. It did have a kick but, on top of that, when it was fired, it went off with a sickening jolt that seemed to make every bone in your body start vibrating.

When it came to my turn, a strange thing happened. In the previous days, we had been firing the 0.303 rifle on the range. The procedure for firing this rifle is to lie down in a prone position with the rifle butt in the shoulder. Then you take the first pressure. What this means is that you take up the slack in the trigger. Then you look through the sights and, when you have the bull's eye aligned, you squeeze the trigger. When I assumed the prone position with the PIAT, I automatically took the first pressure. The problem was that the PIAT didn't have a first pressure. When I pulled the trigger, it fired before I had time to properly sight. Amazingly, it hit the target. We were using dummy shells. That is, they didn't contain explosive. However, the shell was quite large and its momentum was sufficient to leave the drum severely crumpled. This necessitated volunteers to go and prop up the drum with sticks. Everyone continued to fire but no one could hit the target. When it came to my second turn, although I hadn't time to properly sight the first time, I had managed to look through the sights. I remembered that it was sighted on a point about two feet above the target and a bit to the right. I therefore aimed at what I recollected was that point. Again it hit the target. This time, the drum was so severely crumpled that it was not possible to further use it as a target.

We therefore merged into one line and continued firing, no one being able to hit the target. It came to my third turn. Two feet up to the right. This time, the second drum was obliterated. There was no chance of using it again as a target so it was decided to finish up at that stage. Those who were lined up for their third time were relieved. They didn't want to fire this thing again. We learned later that the NCOs (corporal and sergeant instructors) got into trouble for using the PIAT. It was supposed to be "US". I don't know what that meant. Perhaps it meant unsafe. Everything in the military is jargon but you wouldn't dream of showing your ignorance by questioning what any of it meant. Before that time, hardly anyone knew me. But after that, when I walked around the station, I would be pointed out. "That's the bloke who hit the target with the PIAT three times and no one else hit it once".

On the weekends, we could usually go on leave. On one occasion, I got a ride with a friend on his motorbike to go to the city (Perth). To prevent my cap blowing off, I stuffed it into an epaulette. Unfortunately, it must have blown off and I had to go to the store to get a replacement. The sergeant at the store wanted to impress two ladies who were on the staff with his power. When I explained what had happened, he severely reprimanded me and asked, "Where should the cap be?" Of course, I had to answer "On the head". Several years later, when I was a cadet officer in the university squadron, we spent a week's camp at the same air force station. One of our activities was clay pigeon shooting for which we used a double-barrel shotgun. The same sergeant who had upbraided me for losing my cap was there. This time, his job was to run around, out of breath, to pick up the clay pigeons and bring them back. How the mighty had fallen!

On one occasion, the exercise was to start at the top of a sand hill and storm a post at the bottom of the hill. While the attack was in progress, a machine gun (Bren) was set up to fire rounds over our heads to make it more realistic. We carried out the exercise several times but, each time, our instructors were not satisfied and made us repeat it. A friend and I thought it was getting boring so we decided to sit the next one out. We sat down to smoke a cigarette. Then there was a ping and a shower of sand sprayed up between the two of us. It took us a while to realize what had happened. When we did, we thought perhaps we should join in after all. When we arrived at the bottom of the hill, the corporal who was in charge of firing the gun was sharing a joke with everyone. The barrel of the gun that had been set up to fire automatically had accidently dropped and fired a round into the side of the hill. My friend and I joined in the laughter. "Yes, that was really funny", we agreed.

University Days

My time at the University of Western Australia was a happy one. I lived in a university hostel which housed about 100 male students. I enjoyed the camaraderie. It was advantageous to be able to discuss course work with other students and to participate in inter-college sports. In my later years, I was able to help other students in their early years and this was a rewarding experience. Before beginning my science degree, I had the illusion that when I would graduate, I would be a sage and know everything there is to know about science. In fact, the time passed much more quickly than I expected. The time to complete the bachelor degree was less than three years since the academic year began in March and finished in November, just a duration of two years and eight months. I realized that the expectation I held at the beginning was hopelessly ambitious. I was reminded of an apt quotation by Will Durant, an American writer and philosopher. He wrote, "Sixty years ago, I knew everything. Now I know nothing. Education is a progressive discovery of our own ignorance".

I will just recount a few anecdotes that occurred during my fourth year in which I completed an Honours degree in physical chemistry. The degree entailed carrying out a research project. Sometimes, I and some of my fellow students worked back late at night to complete experiments. On one occasion, I didn't finish until late. All the lights in the area were switched off at midnight. I had to return to the hostel by walking through a park before crossing the highway. On this occasion, it was pitch black as I gingerly felt my way through the park. Suddenly, I was hit with a thumping blow on the back that knocked me to the ground. For what seemed a long time but was probably only a few seconds, I was grappling with my assailant. What had happened was that I had passed a water sprinkler, one with long arms that rotated in the wind. That was what had dealt me the blow in the back and I had finished up getting into a tangle with it.

One of my fellow students was a German whose name was Gunter. He had actually been in the Hitler youth but was a very pleasant person, if a little eccentric. The chemistry building had three floors. The ground floor was devoted to physical chemistry, the first floor to inorganic and the top floor to organic chemistry. One day when we were all working in the laboratory, we heard a huge explosion which caused us some alarm. However, one of my fellow students said, "Don't worry, it's nothing. It's just Gunter".

Gunter worked in organic chemistry so he was on the top floor. In his lab, he had a big drum which he used as a rubbish bin. He would throw all his scrap papers into it as well as solvent residues and other waste materials from his experiments. He was also a smoker. Occasionally, he would toss a cigarette butt into the bin and an explosion would result. In those days, we didn't have the strict health and safety regulations that are in place today.

Another anecdote involving Gunter occurred during an exposition that students used to put on annually for the public. This usually began on a Friday evening and continued on into the Saturday. We would set up interesting demonstrations illustrating scientific principles. Gunter had prepared a most elaborate and complex experiment. It was to illustrate automatic titration, a procedure for analysing the concentrations of aqueous solutions. These carefully calibrated solutions were contained in a number of huge glass containers and there were pH meters and burettes incorporated in quite a complicated set-up. I was setting up my demonstration which involved silver and gold electroplating of different objects. Gunter was on the opposite side of the bench. He had set up a burette and needed to raise the level of liquid up to the zero mark. He asked me if I would blow into some tubing so as to raise the liquid level. "Keep blowing", he was saying until he abruptly shouted "Stop!" The solution that I had been blowing into turned out to be a calcium salt solution, the concentration of which he had precisely measured. Every chemistry student knows that if you want to test if a gas is carbon dioxide, you bubble it into a calcium salt solution (lime water) and if it turns milky, this confirms that the gas is carbon dioxide. The large container of liquid that I had been blowing through had turned into milk. What a disaster! "Why did you blow into it?", Gunter shouted. "I blew into it because you asked me to Gunter. I didn't know what it was", I answered.

What followed was something of a spectacle. Gunter stood in the centre of the lab and began bellowing out what were presumably oaths in German. Then he would pause motionless for a short time before launching into his next tirade. The only word I understood was "Scheisen" which he used to begin each soliloquy. All my fellow students found it hard to conceal their mirth. For me, it was not very funny at the time because I had been part of the joke. Just then, some wide-eyed boys rushed into the lab, no doubt expecting to see some of the wonders of modern science. Most of us had not finished setting up our demonstrations. With nothing much to see, the boys formed a circle around Gunter, looking up at him as he projected his wrath and wondering what it was all about. They possibly thought that this must be part of the exhibition.

Another interesting colleague during my time as a student at the University of Western Australia was a young man named Henry. Henry was a bit of a wag but was very bright and was awarded first-class honours in physical chemistry, which was relatively rare at this university. My first recollection of him was when we were waiting for a class to begin and Henry was smoking. The professor walked in and said, "Could I ask you to put out that cigarette?" On another occasion before a class, it was winter and quite cold. The students would stay in the sun until the last moment before entering the building. Frequently, this meant there would be a delay while the stragglers were seated. The professor became tired of this. One day, when the throng of students were filing in, he took hold of the two sides of the door and tried to close them. There was a throng of students trying to push in with Henry at the front. Thinking this was a student trying to play a trick, Henry called out "OK, who's the smart bastard?" The "smart bastard" was the professor who was pushing against the crowd of students. After a few seconds, the force of the students prevailed and they managed to partially open the door. Henry found himself face to face with the professor. Actually, it was more face to chest as the professor was a big man and Henry was small. For a moment frozen in time, there was a stand-off with the students trying to push in and the professor resisting. It was a bit like a tug-of-war except it was a push rather than a tug. Eventually, the message got through to the rear ranks of the students and the professor was able to close the door. Then all the students trooped up the stairs and came into the sloping lecture theatre through the back door. It took about ten minutes before everyone was seated. The professor simply noted that the start time for the class was 11 am. The statement had been clearly made.

In our honours year, I was one of about seven who acted as demonstrators. This involved supervision of first-year students in their practical classes. After finishing, we would get together in a lab to relax by having a competition using a table tennis bat and ball. The game would start with a simple exercise of hitting the ball ten times with each hand. Then the procedure would be progressively made more difficult, ten times alternately in front and behind, ten times against the corner of the room, ten times against the ceiling and finally, the "piece de resistance", going around the room hitting the ball against successive walls. This was the most difficult of all and I don't think anyone managed to complete it. The lab had a glass door at the front. While one of us was competing, the rest would sit on chairs at the back facing the glass door. The lab had a bench with several intricate beam balances that were very delicate. To use them, you had to follow strict rules – never touch them, always ensure that the samples being weighed were at room temperature, use pincers to place the weights on the balance pans, etc. I was terrified of these balances and always gave them a wide berth. On one occasion, Henry managed to reach the final step of the game, hitting the ball against alternate walls. He was flailing the bat, smashing the balances, bits of the balances were falling off and sent flying in all directions. At that moment, the frame of the professor appeared outside silhouetted in the glass door. All of us who were watching were paralysed. We couldn't warn Henry and, even if we did, he wouldn't have taken any notice, so engrossed was he in the task. The following day, a notice was sent around to say that there were no further games to be played in the balance room.

Henry's project for his honours degree was to measure the ultraviolet spectrum of benzene. Prior to doing the measurements, he had to purify the benzene. Many of the published spectra were made with what was thought to be impure benzene. One of the impurities was water and this was removed by successive extractions of the benzene with concentrated sulphuric acid. Henry had managed to get his benzene to a very pure state but decided to do a final extraction with sulphuric acid just to make sure. While he was preparing the final step, someone "inadvertently" took the bottle of concentrated sulphuric acid that was on the bench and replaced it with a bottle of concentrated nitric acid. Unlike sulphuric acid, nitric acid reacts with benzene. Henry's pure benzene had been converted to nitrobenzene.

Vacation Work

During my four years as a student at the University of Western Australia, I spent three vacations working in different jobs which were necessary to supplement my Commonwealth Scholarship allowance. I will briefly summarize the jobs I undertook. After my first year, I took a job on a wheat farm. I took a train from Perth which arrived at the station just after midnight. I was met by the farmer who took me to his property where I bunked in an outhouse. At about 6.00 am in the morning, he woke me to begin my first day's work. Our first job was to load a truck with bags of wheat that had been harvested the previous day. To get the bags on the truck, my job was to drag them to the truck where a hydraulic bag lifter was placed. The bag lifter was made by Horwood Bagshaw, an Australian manufacturer of agricultural machinery, at a time when Australia had a strong manufacturing industry. A lever was pulled and the bag was lifted on to the truck where the farmer put it into position. It sounds easy but the platform on which the bag had to be placed was a couple of inches from the ground. After my first day, I lost most of the skin from my hands due to blisters from lifting the bags on to the bag lifter. I only realized afterwards that the secret was to use your leg for leverage to help lift the bag on to the platform. After loading all the bags, we went to the silo via a weighbridge and emptied them into the grain bin. The weight of the grain was calculated from the difference in weight of the truck before and after leaving the silo.

We then returned to the house for breakfast. The farmer's wife always prepared a substantial breakfast of cereal and usually steak and eggs with toast and copious cups of coffee to quench our thirst. After breakfast, we returned to the field. My job was relatively easy. I just had to drive the tractor around the field while the farmer sat on the harvester, ensuring that there was no clogging and regularly stopping to fill the bags with grain. We would take bottles full of tea and leave them near a tree. The heat was such that the tea was kept at the right temperature.

Prior to recruiting me, the farmer told me that he had several men to help him but none could do the work and he had to terminate them. He seemed quite pleased with me as I had experience working on a farm (although not a wheat farm) and I knew how to drive. There was one anecdote that I will recount. It was close to the festive season and a party was arranged for all the farmers in the neighbourhood and their families. A number of kegs of beer were at the ready. Everyone was lined up, all dying of thirst because it was very hot. When the tap of the first keg was opened, everyone expected to see the lovely beer flow but all that came out was foam. There was a mechanism for priming the keg to facilitate the flow but the person in charge evidently had been too enthusiastic with the pump so that the beer had been converted to foam.

On another occasion, the farmer's wife came running out to where we were working and was in an upset state. All their chickens were suffering from heat exhaustion and many had died. They may not have had sufficient water. Although my task of driving the tractor was fairly easy, the heat was oppressive. As I drove around, occasionally I would receive a spray of water from the tractor radiator and I always looked forward to this because of its wonderful cooling effect.

During the vacation following my second year at university, I worked in two jobs. My first job was as a door-to-door salesman trying to sell subscriptions for a Sunday newspaper. I did have some success but it was far from being an outstanding performance. It did, however, help me to appreciate the difficulties involved in salesmanship and to learn how to accept a large proportion of refusals. My second job was as a barman in a hotel. The hotel had a very long bar and, at the busiest periods, there were three who attended customers. I coped pretty well. The drinks were mostly beer served in different sized glasses. I soon learned the prices of the different drinks and had no trouble serving and giving back the change. If I were to admit a regret, it was that, although I worked efficiently, I didn't engage personally very much with the customers. Of course, I was young then and hadn't developed the empathy that one should ideally have when dealing with the public.

On one occasion, a customer ordered drinks but carried them away without paying. The other bartenders noticed this. When we tallied the takings at the end of the day, there was a deficit. The hotel manager told me to try to get the money from the man who hadn't paid. The following evening when the same man bought drinks, I gave him the change, explaining that I had withheld the money he had "forgotten" to pay the previous day. He began to argue aggressively. The bartender who was working next to me was experienced and was aware of what was happening. He came over and suggested to the man that he might like to move on, which he promptly did.

After finishing my third year and obtaining my B.Sc., I procured a job in which I had to set up a small laboratory. My employer ran a company that produced fertilizer from abattoir waste. My task was to try to optimize the processing conditions to obtain the highest quality product. The plant was in an outer suburb of Perth and about three miles from the commercial centre. On my first day, I took a bus and banked on finding a business that rented bicycles. By a stroke of fortune, there was one and I was able to cycle to the plant. My boss provided the equipment needed to carry out the tests and I set it up in what was really just a shed. It was mid-summer so it was very hot and the heat was amplified by some of the procedures I used such as moisture determination which involved heating.

Reflections of My Early Days

Thinking back, I have had many experiences where I feel debts of gratitude and I believe it is important to acknowledge them. My early days subjected me to testing situations. Although they may have been tough, they were well within my powers to cope. They enabled me to develop resilience and self-confidence. Many young people these days are deprived of these experiences. Life is made too comfortable and risk-free. This message is portrayed in the book "The Coddling of the American Mind. How good intentions and bad ideas are setting up a generation for failure (Lukianoff and Haidt, 2019)". The authors explain how this unfortunate situation is brought on by social trends such as fearful parenting, decline of child-driven play and the emergence of social media. Another experience that I felt was important for developing fortitude was military training, first in school cadets and then in National Service. For a time after 1911, participation in the Cadet Corps was compulsory for all males in Australia. Nowadays, it is no longer compulsory although it is carried on in some schools such as Kings School which is located in Parramatta, New South Wales. Military training for men who reached the age of 18, known as National Service, was compulsory for a short time in Australia and involved times of three to six months depending on the Service. This was phased out in the 1970s. In many countries, military service is mandatory. In Israel, both men and women participate.

The other area where I feel a deep debt of gratitude is the education system in Australia and to the governments and taxpayers who have supported it. I consider myself extremely fortunate to have been assisted throughout my entire education except for my first year at university. First, I was awarded a secondary school scholarship. Then, after my first year as an undergraduate at the University of Western Australia, I was awarded a Commonwealth Scholarship which enabled me to complete my Bachelor of Science degree there. Then, I received a postgraduate studentship (first given by CSIRO and later maintained by the University of Sydney) that made it possible to obtain a Ph.D. degree from the University of Sydney.

I have mentioned the gratitude that I feel for the educational opportunities that I was given, thanks to the sense of responsibility and vision by different governments and educational institutes in Australia. Apart from this, how fortunate I have been to live in a first-world country and prosper from its benefits! I realize that this opportunity has been handed to me by those who have come before and whose hard work and sacrifice have made it possible. I therefore find it hard to understand the ingratitude of some members of our younger generation. They don't seem to appreciate how fortunate they are. Instead, some develop a resentment towards our culture. They appear to look for faults rather than see the great things that have been achieved. I feel that this attitude has been inculcated by their mentors. Their minds seem to have been influenced in a negative way by the present education system. The positive aspects and pride in the accomplishments of the nation seem to have been eliminated from curricula and negative ideas have been planted in their malleable minds.

What Mistakes Have Been Made?

Until now, I have given well-deserved kudos to governments. It is therefore with some regret that I have to point out what I feel have been some of the flaws in government decisions that have led to detrimental effects on society even though they may have been well intentioned.

The first mistake that I believe has been made is to allow phasing out of military training. I was fortunate to have had training, first in the School Cadets, then in National Service and a short time as a cadet officer in the university squadron. I see two important benefits that arise from military training for young citizens. One is that it contributes to character building by instilling self-confidence and the capacity for physical and mental endurance. These are qualities that may not be developed in the comparatively easy lifestyle that many young people currently have. I have heard young men say that they oppose military training because it involves taking orders and they do not take orders from anyone. I think this is a naïve and misguided attitude. In an armed conflict, it is essential to have a chain of command. The same thing applies in team sports and in many workplaces. Teamwork counts and there usually needs to be leaders to make it work. It also should be understood that an important component of discipline is self-discipline.

The other benefit is naturally that training provides the preparation that is needed in the event of armed conflict. Many people today are complacent. They don't visualize that their country could face future danger. Most young people have never lived through wartime and assume that peace will be preserved indefinitely. History tells us otherwise. Australia is an attractive country because of its large land mass and resources. The world is not made up entirely of friendly and peaceful peoples. There is envy and resentment and thirst for power. The best policy to ensure that it is not forcibly taken over is to project an image that it has a potentially strong military force and this will act as a deterrent. The decisions to neglect military training by politicians have not been in the best national interest.

Another area where the country has been let down by its leaders is in the education system. We will only refer here to decisions made by the parliament in relation to tertiary education and topics such as the school curriculum will be discussed later. As I mentioned, I was fortunate to have accessed a good system of student assistance. The Commonwealth Scholarship Scheme was introduced by the government in 1951 some years before I began university (Daniels, 2017). In 1951, 3,000 Commonwealth scholarships were awarded and selection of recipients was made on the basis of academic merit. The objective of the scheme was to promote participation of the most capable students rather than encouraging broader participation in tertiary education.

The rationale for award of student assistance was changed by the government in 1974. The new scheme enabled all full-time students to receive assistance provided they qualified under a means test. This meant that the objective now was to broaden participation while removing the competitive nature of the earlier scheme. It resulted in a steep increase of student numbers. It is interesting to compare student numbers over time. In 1949, Australia's largest university (University of Sydney) had 4,500 students. In 2003, the largest university (Monash University) had a student population of 48,477. In 1949, the number of students enrolled in Australian universities was 31,753 compared to 828,871 in 2003. Of course, we have to take into account other changes such as the country's larger population and the influx of a greater proportion of female and foreign students. Nevertheless, tertiary students comprised 0.2% of the population in 1939 compared to 3.3% in 2001 (Abbott and Doucouliagos, 2003).

The changes introduced by the government in 1974 have caused an explosion in the number of students taking advantage of the system. It inevitably led to a blow-out in costs which the country could not sustain. A Higher Education Contributions Scheme (HECS) was brought in from 1989 in which fees for university study were gradually re-introduced. Students were allowed to take out loans and defer the debt until their income after graduation exceeded a threshold level. The HECS scheme has had a number of makeovers during the years since its inception. The loan scheme, now called Higher Education Loan Program (HELP), has seen the average debt steadily increase at a rate greater than inflation. In 2011/2012, 2.1% of those with loans were facing debts of more than $50,000 and by 2015/2016 this had increased to 6%. The total HELP debt to the Commonwealth was $25.5 billion in 2011/2012 and this had increased to $62 billion in 2017/2018.

As I look back over the past 70 years, which decisions about university funding seem to have been positive and which have been negative? The introduction of Commonwealth scholarships in 1951 seems to have been positive. The selection of recipients was based on merit. A programme, which was costed and government could afford, apportioned scholarships to students based on their ability and capacity for study. In 1974, a decision was made to nominally make university education accessible to all who wished to avail themselves of the opportunity. At first thought, this seemed to be a noble gesture. The result would be the creation of a better educated cohort of citizens. That should, in theory, enhance the culture and prosperity of the nation, a positive outcome. Some of those who benefited from the new system have claimed that it gave them an opportunity to gain a profession which otherwise would not have been possible. Some prominent people were among this group, notably celebrities who praised the government decision and, being celebrities, were featured in news items. However, their claims were not wholly true for which my experience can testify. I entered university before 1974 and was able to do this by working for several years to save sufficient money to pay my expenses for the first year. After my first year, I obtained a Commonwealth Scholarship. This allowed me to continue university, providing I succeeded in passing exams each year, which I was able to do.

What have been some of the negative outcomes? Prior to 1974, students who chose to enter university were more likely to be aspirational and goal-oriented. When conditions are made easy, this tends to diminish the proportion of students having the motivation needed to succeed. It has been evident in the increase in drop-out rates that have followed. The surge in the numbers of tertiary students resulted in the introduction of HECS fees. The debt incurred that had to be repaid has imposed a financial burden after graduation. The unpaid debts of those who failed to complete their courses are imposts on the government and ultimately on the taxpayers. The problems have been succinctly summarized by George Binning (2019) who wrote, "Universities have been admitting too many students, a greater proportion are now struggling, their debts have soared, and the payoff for those who graduate has waned according to the Productivity Commission". The reliance of Universities on fees from foreign students is raising questions about the risk of prioritizing income from fees to the detriment of maintaining standards. My overall impression is that the system of educational assistance was a good one when I was a student. The changes that have been introduced since then, although seeming to be magnanimous and were applauded by many at the time, have had unintended consequences that have not always led to positive outcomes. The old adage "If it's not broke, don't fix it" may well apply. In recent times, Universities have become more dependent on income from foreign students in order to survive. This dependence is raising questions about the maintenance of academic standards and whether the welfare of Australian students is being given due priority.

References

Abbott, M. and Doucouliagos, C. 2003. The efficiency of Australian Universities: a data development analysis. Economics of Education Review 22: 89–97.

Binning, G. 2019. Universities have been admitting too many students. Spectator Australia, July 5. (www.specator.com.au>author>george-bunning).

Daniels, D. 2017. Student income support: a chronology. Australia Parliamentary Library, Canberra, August 1.

Lukianoff, G. and Haidt, J. 2019. The Coddling of the American Mind. How good intentions and bad ideas are setting up a generation for failure. Penguin Press, London.

chapter two

Scientific Thinking and How It Can Be Applied

There is a saying that "history is about things that didn't happen written by people who weren't there". In this chapter, I will be trying to avoid this criticism by attempting as much as possible to describe events that did happen and I was there to witness them. My background is in science. Those of us who have specialized in some area of knowledge have the obligation to convey aspects of that knowledge in an easily assimilated form. In that way, we may be able to enrich the lives of our fellow travellers and those who follow us. That is what our ancestors have done to make our lives better than they would have been. I will therefore try to outline what I believe are some of the important things to have come out of science. That doesn't mean that I will be sending the reader off to peruse Albert Einstein's papers on the theory of general relativity. It will be a lot simpler than that. The philosophy of the scientific method will be detailed as we go along. To make a start, one important aspect that can be expressed in simple terms is that knowledge is advanced by making mistakes and learning from them. Of course, this attribute is not restricted to science although it is one of its basic foundations. There is a story about a man who went through his whole life without ever making a mistake. Another feature of this special man is that he went through his whole life without ever learning anything. Of course, this is an exaggeration. Everyone learns something during their lives, even if it is only how to tie their shoelaces. The real question is whether it was something of value or whether it was to clutter up their mind with useless or false information. Much of what pervades the media these days is just that, such things as black lives matter, gender fluidity, or catastrophic climate change.

Early Research Experience

Following graduation from the University of Western Australia (B.Sc. Hons), I moved to the University of Sydney. The reason for the move was simply that I was offered a Commonwealth Scientific and Industrial Research Organization (CSIRO) studentship to work with Professor A.E. Alexander in the Physical Chemistry Department there. Alexander had a

DOI: 10.1201/9781003254065-2

group of graduate students working in the field of surface chemistry and colloids. My project was to study the behaviour of proteins at surfaces (the more general term is interfaces). Surface chemistry involves the study of monomolecular films; i.e. films one-molecule thick, a unique state of matter. These films, called monolayers, have a thickness of roughly between 10^{-6} and 10^{-7} cm. Of course, this doesn't mean much to any of us except to suggest that their thickness is very small. Films that are one-molecule thick are called insoluble monolayers. They are called insoluble because they do not dissolve in the subphase (usually water). They cannot be seen by the most powerful optical microscope but can be manipulated and their properties accurately measured. For example, monolayers change surface tension. The surface tension of pure water is approximately 72 mN m^{-1} at room temperature. Monolayers can reduce this value by up to 40 mN m^{-1} or more and the changes can be measured easily to 0.1 mN m^{-1}. The change in surface tension is referred to as surface pressure since it is a two-dimensional (2-D) analogue of ordinary three-dimensional (3-D) pressure. The units of ordinary 3-D pressure are mN m^{-2} whereas the units of 2-D pressure are mN m^{-1}. This reflects the loss of one dimension when we move from a 3-D to a 2-D system.

Monolayers can be studied using an instrument called a surface balance. In this instrument, a measured amount of a compound can be spread at a surface as a monolayer on a trough at the air/water interface. The monolayer can then be compressed (or expanded) using impermeable barriers and properties such as surface pressure can be monitored as a function of the area occupied per molecule. The resulting relationships are called 2-D pressure-area isotherms. In addition to surface pressure, other properties such as surface electrical potential, surface viscosity and elasticity, and optical properties can be measured as a function of area per molecule. Most of the properties that are measured in 3-D systems have their analogues in 2-D.

Compounds that form stable monolayers have molecules that possess a dual character. They have a hydrophilic (water attracting) and a hydrophobic (water repelling) group. Such compounds, which are called surface active, can be stable at an air/water interface because the hydrophilic group can interact with the water and the hydrophobic part can escape from the water. As a result, molecules are orientated and can be highly concentrated. Compounds that form monolayers play an important role in some physical and chemical systems such as in the stability of foams and emulsions. For example, a stable foam cannot be formed by bubbling air into water. A few stray bubbles may appear but this is caused by impurities in the water. If pure water is used, the bubbles burst immediately. Stable bubbles can only be created when a surface-active compound is present to form a film at the air/aqueous interface. Only small amounts of the compound are needed since it only requires sufficient to form a mono-layer at the bubble surface.

The Role of Dimensions

I have included this brief description of the research area because 2-D chemistry is a fascinating field and may be of interest to those for whom it may not be familiar. In fact, the concept of dimensions plays an important role in science. There is a story about a group of flat-earthers who set out on a journey. They started from a point and walked in a straight line towards the horizon. They were stepping carefully because they didn't want to suddenly come to the edge and drop off. They walked for miles, in fact for tens of thousands of miles, but never came to the edge. They did, however, come to a place that seemed very familiar. It was the point from where they started. How weird was that?

Something similar happens if you set off to try to find the edge of the universe. Of course, you wouldn't walk. The galaxy that we are in has some billions of stars and is shaped like a disc. That is why when we look out, we see regions where there are few stars but then there is a bright band of stars stretching across the sky. This is because we are inside the disc looking towards its rim. We call this the Milky Way and call our galaxy the Milky Way galaxy. The long axis of the disc (our galaxy) is about one hundred thousand light years from rim to rim. A light year is not a time, it's a distance. It's the distance light travels in a year. The speed of light is about 186,000 miles per second. That means that, if we got into a spaceship that travelled at the speed of light, it would take us about one hundred thousand years to cross from one rim of our galaxy to the opposite rim. Of course, if we were smart, we would cross at the small axis of the disc. This is only about one thousand light years across and it would just take about a thousand years, so it pays to be smart.

If we managed to escape from our galaxy, we would find that there are billions of other galaxies outside. However, when we travel at such high speeds and traverse huge distances, we need to give up the simple concept of 3-D space. Understanding questions about the universe requires adding the dimension of time so that we have to think in terms of a four-dimensional space-time continuum.

Discussion of dimensions serves as an introduction to the topic that I wish to develop in this chapter. The topic is the scientific method and how it is applied in research. Most people think of science as including observation and experimentation and this is true. It does, however, involve more than just that. A knowledge of the method not only helps to understand how science works but also contains concepts that can be valuable for non-scientists in their thinking about general problems that they confront.

Learning an Important Lesson

A short time after beginning my research project, I came up with a new theory to explain a specific behaviour of proteins at a surface. Strictly, it should be called a hypothesis. I will stick to the word theory for the present because some of us Australians, especially scientists, have trouble with words of more than one or two syllables. I outlined the theory to my supervisor. He listened patiently and suggested some experiments that I could do to check if the theory was right. At the time, I thought it was unnecessary to do the experiments as I felt sure that my interpretation was correct. Nevertheless, on reflection, I thought that I should do the experiments. Then, when I had finished, I would barge into my supervisor's office, throw the results on his desk and say "I told you so". Of course, I wouldn't have done that. I was pretty arrogant in those days but not quite that arrogant.

The only problem was that when I came to do the experiments, they did not turn out at all as I had expected. In fact, they showed that my theory couldn't possibly be right. I was shattered. I had felt so sure that this was the correct way of looking at the problem. I couldn't return to the lab for several days. When I did manage to drag myself back, it took some time before I could refocus. I had to eat big helpings of humble pie and it tasted horrible.

However, when I returned to wrestle with the problem anew, I began thinking along a different line. I found that my new line of thought was better and was more soundly based than the one I had held initially. This experience was to be repeated several times during my early time as a research scientist. After a while, I realized that the humble pie didn't taste so bad. In fact, eventually I began to like it. What I realized was that each time I formulated a theory, carried out experiments to test it, and showed the theory to be false, I was able to move on and come up with a better and sounder way of looking at the problem.

The Scientific Method

It was only some years later when I began reading up on the philosophy of science and particularly the writings of Karl Popper (Popper, 2002) that I realized that I had stumbled on the scientific method. I had obtained a degree in science and begun a career in scientific research but I had received no instruction about what constituted the scientific method and didn't properly understand what it was. It was really deceptively simple, at least in principle. It begins with a conjecture or hypothesis about how to explain some observations. Experiments are then designed to test the hypothesis. The essential requirement of a scientific hypothesis is that it must be refutable. This is what distinguishes science from non-science or pseudo-science. If the experiments refute the theory, it is rejected and thought given to a new theory. The new theory can make use of what was learned from the refuted theory. If severe tests are applied to the theory and it is not refuted, we can say that it has been corroborated. It does not mean that we have arrived at the absolute truth but, at least, we can be fairly confident that we are moving towards truth. A hypothesis or theory in science is never assumed to have been proven. This doesn't seem to be understood by many who engage in public debate. An obvious example that is often expressed is that the science of climate change has been "settled". This is erroneous. The science is never settled. Science proceeds by a trial and error procedure in which scientists admit their mistakes and learn from them. This enables them to come closer to the truth but never to claim that they have arrived at an absolute truth.

Of course, if we are only able to propose theories, subject them to tests and refute them, this doesn't get us very far. It does get us a little way. At least it eliminates one possible theory and may help us to advance towards a better one. However, if science is to progress, we also need to have successes. This means that when we throw everything we have to try to refute a theory and we are unable to refute it, this gives us some confidence that we may be on the right track. It doesn't mean that we have arrived at the absolute truth. Nevertheless, we are probably justified in saying that we are moving in the direction of truth.

Karl Popper illustrated the criterion of refutability by comparing a number of theories that were being discussed in the early 20th century. These included the two psychological theories of Sigmund Freud and Alfred Adler, the theory of history of Karl Marx, and the theory of relativity of Albert Einstein. Neither of the two psychological theories could be considered to be scientific because they were so general that it was not possible to devise an experiment that could refute them. This property of irrefutability was held up by their adherents as a proof of their veracity. For Popper, it meant just the opposite. The theories could not be accepted as scientific. In a sense, they were similar to the predictions of fortune tellers. By making their predictions so general that they become highly probable, it is impossible to refute them. The theory of Marx was refutable but Popper asserted that it had been tested and was refuted. However, its supporters have introduced auxiliary hypotheses to make the theory and the evidence agree. This can always be done but it results in a loss of scientific status of the theory and can lead to the theory becoming unfalsifiable and therefore unscientific.

The theory of Einstein was different. One of the predictions of the theory of relativity was that light would be deflected by gravitation. This was tested in 1919 by photographing the positions of distant stars which became visible when their light passed near the sun during a total eclipse and comparing their positions in the night sky. The experiment showed that the light was deflected and the direction and magnitude of the deflection was within experimental error of what had been predicted from the theory of relativity. It therefore satisfied the criterion for a scientific theory; i.e. that it was refutable. If the light had not been deflected or even if the direction or magnitude of the deflection had been different to what was predicted, the theory would have been refuted. Karl Popper produced several books but some are written more for philosophers. The book that I believe is most readable for laypeople is "Conjectures and Refutations" (Popper, 2002).

The Generality of Applying the Scientific Method

The way of thinking inherent in the scientific method shouldn't only apply to science. Everyone is a scientist. We all studied science at school and we follow and understand many of its applications. We are all accountants. We balance our budgets at home, or try to. We are all lawyers to some degree. We understand a lot about the justice system and how it operates. Many of us are plumbers. We replace the washers in our taps, usually several months after they start to drip, but we do it. The point I would like to make is that all of us can make use of knowledge from different professions and trades to enrich our lives. By enriching our lives, we enrich our society. You don't need to be a mathematician to understand that the total richness of a society is equal to the sum of the contributions of its members. In view of that, I am going to give some suggestions of how we might all apply the scientific method to enhance the quality of our thinking.

First, we should never hold a dogmatic opinion. All of us bring our baggage when we come to discuss issues. This includes all the biases and prejudices we have formed from our experiences. We need to try to leave this baggage aside and see the issue from the perspective of a detached observer. This is not easy to do. It requires constant practice so that it becomes instinctive. We need to recognize that the information we receive from a given source may not always be the truth. I think that I am the most agreeable person one could wish to meet. If you ask my wife, she would probably say that I am the most stubborn person she has ever encountered. It is always best to get a second opinion and this especially applies to opinions about oneself. Opinions by others are often more objective. Ideally, even a second opinion is not sufficient. It is preferable to seek information from multiple sources.

Recently, I spoke with someone whose views I found rather strange. When I asked where he had got the information on which he based his views, he replied that it was from multiple sources. However, when he revealed his sources, it turned out that they were all from the same part of the political spectrum. In order to arrive at authentic opinions, it is vital to explore a wide range of viewpoints, even extreme ones: in fact, especially extreme ones, because history has shown us that these can be the ones that are most dangerous to society. Then we need to critically examine each one before forming an opinion. This opinion should in any case be only provisional if we are applying the scientific method. We must never form an inflexible opinion. We should always be ready to change or at least modify it in response to new facts or rational arguments. Frequently, we meet people who, when they speak, it seems obvious that they are getting their information from only one source. It may be a radio station, a TV channel, or a newspaper and it is sometimes possible to guess which one because their views are aligned with a source that has a specifically biased way of thinking. It is important to free ourselves from holding a dogmatic opinion. There is a message being spread that the world will end in 10 years (or is it 12?) if something is not done to address the problem of global warming. To a rational person, this sounds like nonsense. But suppose the world does end in ten years. It could, we don't know for sure that it won't. If it does, many of us will have egg on our faces although that won't matter because we will all be gone. The point to make is that we should never adopt rigid beliefs, even about what we consider to be the most unlikely events.

Another important capability is to be able to readily admit being wrong and to learn from our mistakes. One of the most meaningful lessons I have learned and possibly the most meaningful from my scientific experiences is to be able to recognize that, after carrying out tests, my theory could not be correct. As I mentioned earlier, this invariably led to a sounder and better way of looking at the problem. I have observed that some of my scientific colleagues are often reluctant to give up their theories or ways of thinking even when some results do not accord. This reluctance is understandable. After all, when someone proposes a theory, this is a creative step. It may be that they were the only one in the world to have thought of this idea. Perhaps a few results are at odds with the theory. These can be swept under the carpet; out of sight is out of mind. The important thing for them is not so much that the theory may be wrong but it was their creation and they feel a certain ownership of it. This is a dangerous path to take. It will mean that they become stuck in a rut and are not able to escape from it. If they would give up their theory or direction of thinking, this could lead to a fresh vision that could open up new and more valuable perspectives. The facility to readily admit errors becomes habitual. It is an attribute that I believe needs to be developed not only for successful scientists but for everyone. When Albert Einstein arrived to take a position at Princeton University, he was asked what he would need for his office. His reply was "a desk, some pads, a pencil and a large wastebasket to hold all my mistakes". Some of the most valuable insights into mistaken beliefs are provided by those who have held a certain conviction when they were younger and have realized that they had been wrong. Peter Hitchens, a noted British speaker and author, was a Bolshevik in his younger days. Later in life, he discarded this ideology. The result of having lived among members of the group helped him to understand how their minds worked. This knowledge was valuable for his later thinking and for passing the knowledge on to his peers. Another prominent writer who changed her allegiance later in her career is Melanie Phillips. She held a senior position at the Guardian newspaper which is left leaning. She initially felt that her world view was the same as her colleagues but later realized that this was not so. Her experiences are related in her book "Guardian Angel. My Journey from Leftism to Sanity".

Another skill that we all need to master and which is really an extension of the two previous ones is to be always prepared to change our opinion or at least modify it. When we discuss or debate an issue, we need to listen carefully to what the opponent of our argument is saying. This is not just to be polite or altruistic. Frequently, when we hear the opinion of an opponent in a debate, it can give us an insight into a problem that we have not previously considered. The same thing may happen when we read a passage in a book or hear a view expressed in the media. The more we can make ourselves detached, the more is the likelihood of capturing a useful idea. This will help to re-examine a belief that we held previously. It may also work the other way. If we see a fault in the reasoning of an opposing view, this can give us more confidence that what we had thought is correct. Our thinking always needs to be dynamic. The opposite is to be dogmatic and this is a very bad fault to be burdened with.

Postdoctoral Experience and Employment in CSIRO

After obtaining my Ph.D., I spent five years abroad, the first two and a half years as a research scientist at the Unilever Research Laboratory in Port Sunlight, England. While there, I met a Professor from Chile who was spending a sabbatical at the laboratory. He was from a department of physical chemistry at the University of Chile. He invited me to accept a position in his department as a visiting professor. I then spent another two and a half years at the University of Chile in Santiago. This was a wonderful experience. I had to adapt to a new culture and to learn to give lectures in Spanish. The students who I taught were in senior classes and, when I began, they gave me great support, helping me as I struggled with the language.

I returned to Australia to take a position as a research scientist with the CSIRO of Australia. The position was in the Wheat Research Unit. The CSIRO was structured in a large number of Divisions and smaller groups called Units, distributed throughout the country. Each Division and Unit was dedicated to an Australian industry. When I joined, the most important industries were minerals and wool. There were also Divisions directed to fundamental research such as the Division of Radiophysics and National Standards. At the time, I joined the Wheat Research Unit, 20% of the nation's wheat production was used for domestic consumption and 80% was exported. Thus, it was an important industry for the economy. The exports had to compete with those from the other main exporting countries which at that time were the U.S.A., Canada, and Argentina. Since that time, the percentage of the wheat crop that is exported has decreased slightly but wheat still remains an important export commodity.

CSIRO is publically funded. That is, it is financed mainly by individual taxpayers with a smaller contribution from industry. In view of that, it seems reasonable to expect that its scientists should accept the responsibility to inform the public of their research and to justify its impact on the nation's prosperity. I will therefore spend a little time to describe the work that was done in my Unit and try to do it in a way that laypersons easily understand. As mentioned, our main mission was to improve wheat quality for the benefit of Australian consumers but, also importantly, to be competitive in the export market. This task is shared by a number of different sciences. Wheat breeders use crossing of parents and selection of progeny to target such aspects as grain yield and resistance to diseases. This requires knowledge of genetics. Plant pathologists are needed to solve problems caused by diseases that can severely affect the crop. The CSIRO Wheat Research Unit (later renamed the Grain Quality Research Laboratory to cover other grains besides wheat) was a small group comprising chemists concerned with quality after harvest. Our task was to study the chemical composition of the wheat grain and relate it to its end-use quality. There are a range of products made from wheat. The main ones are Western-style pan bread, flat bread, biscuits (called cookies in North America), pasta, and noodles. The quality requirements for each can be different. For example, Western-style pan bread requires wheat with hard grains that are milled to give flours having strong dough properties whereas soft-grained wheats with weak dough properties are preferred for biscuit manufacture.

The composition of wheat flour is complex and comprises a huge number of distinct chemical compounds. Starches are by far the most abundant general class of compound. However, proteins which usually make up 10–15% of flour are the components that determine many quality aspects such as dough properties. A large number of different proteins occur in a wheat variety and each one is coded by a gene. In simple terms, the aim is to discover the types and proportions of the different proteins that determine the optimum composition required for a given product. Then the genetic information can be used by the breeder to target the protein composition that is required for the product. This is a simplified description of the strategy that was applied in our research on wheat quality. Our work was therefore fundamental and long term in nature.

However, there was also direct applied research being carried out in the Unit. I will describe one example of this. A problem encountered in wheat production is the damage to the wheat grain from rain just prior to or during harvesting. Other types of weather damage can occur such as is caused by frost or heat but the work of the Unit was aimed at moisture from rain. Exposure to rain can cause sprouting. We tend to think of sprouting as the appearance of sprouts emerging from the grains. However, from a scientific point of view, the process of sprouting is understood to begin beforehand. When the wheat grain is exposed to moisture, the process of germination begins well before any visible signs. Chemical processes occur involving the enzymes that are present. The effect of rain-damaged wheat becomes apparent when the grain is milled and even more obvious in bakeries. The dough can become sticky as a result of the enzymes breaking down molecules of starches and proteins, making it difficult to process and giving a poor-quality product. The problem that our Unit was faced with was to devise a method to determine the degree of sprouting. This problem, in which I did not directly participate, was tackled by my colleagues. They developed, in collaboration with an Australian company, Newport Scientific, an instrument which gave a quantitative measure of the degree of sprouting. The principle was based on measuring the viscosity of a suspension of the ground grain. This instrument, the Rapid Visco Analyzer (RVA), is now used throughout the world. Its applications have widened from measuring the degree of sprouting in grain to many others in the food industry such as characterizing different starch samples.

Reflections of Research in CSIRO

I spent 30 years as a research scientist at the CSIRO. Previously, I have commented on how thankful I felt for chances that were given to me in my education. Likewise, I feel a deep gratitude for the opportunity to work for a research institute that had become famous for its high standard of research. Once I heard a colleague, referring to his career in research, say "I have been able to do what I most like doing and I get paid for it". This is a sentiment that I share. Again, I have to recognize that my career depended on the support of the Australian public and the vision of successive governments. Employment as a research scientist working to advance knowledge to benefit the nation must be one of the most fulfilling activities one can aspire to. There are great rewards to being a research scientist besides the intellectual satisfaction. Science is not affected by national boundaries. It gives opportunities for travel to attend conferences, visit research establishments, and to form friendships with scientists from other countries.

What a privilege it has been to work in an acclaimed research institute. There was a change in culture of the organization during my last few years there but this has been covered in earlier books (MacRitchie, 2011, 2018).

In the Preface, I declared that I would try to relate history from the perspective of events that really happened and I was there to witness them. Based on that intention, I would like to reflect on my perception of how scientific research may have changed during my time as an active scientist. This covers a period of three decades as a research scientist at the CSIRO followed by another decade as a Professor at Kansas State University and then a few more years continuing as Editor-in-Chief of a scientific journal.

How Has Scientific Research Changed in Recent Decades?

I shall begin by referring to one aspect of science that was proposed by Karl Popper. This was the need for it to grow or to progress. The growth of science does not mean a continuous accumulation of observations. It means the development of new or improved theoretical concepts that increase our understanding. The way that this occurs is by criticisms of current theories, attempts to overthrow them and, should they be overthrown, imaginative proposals for better theories. The new theories are then tested and subjected to the same critical treatment. This is the fundamental nature of science and the procedure is essential for its growth. It is the hypothetico-deductive method as distinct from the inductive method which simply depends on arbitrary observations. In the hypothetico-deductive method, the experiments that are carried out are not arbitrary but are designed to severely test the hypothesis. As a result, these experiments are ones that would never have been thought of if it hadn't been for the initial theory and the criticisms and attempts to overthrow it. This then is the innate nature of the scientific method and its application is what leads to an increase in our understanding.

What happens if science doesn't grow? Or, is this not a valid question? There seems to be a widely held belief that science will always continue to progress. Certainly, this appears to be the case if we consider the great advances that have been made over the past few hundred years. While there are scientists engaged in research, how could science not be progressing? Popper's criterion of the need to grow suggests an answer to this question.

First, let us reflect on the difference between the inductive method and the hypothetico-deductive method. The difference may be illustrated by a quote from one of the truly great scientists, Albert Einstein. Einstein said, "Imagination is more important than knowledge. For knowledge is limited to all we know and understand, while imagination embraces the whole world and all there ever will be to know and understand". This statement captures the essential difference between the two procedures. The inductive approach can be compared to knowledge. It limits our understanding only to little more than we already know. Hypothesis-deduction, in contrast, has no limits. It allows the imagination to soar into realms that were not previously dreamed of to propose a conceptual idea in an attempt to explain an issue. It then uses a trial-and-error approach to test its validity. This is the way that science can grow and why its growth is an essential requirement for science to remain relevant. If this doesn't happen, then science stagnates and its drive to reveal new and useful knowledge falters.

Now to return to the topic of how science has fared in recent decades. I will restrict my observations to only the narrow field in which I have been involved-cereal science. Journal of Cereal Science has been the premier journal in the field based on its impact factor. The journal was inaugurated in 1983. I was Editor-in-Chief of the journal for one period of just over a decade and was one of the editors for a time prior to that. A journal contains a record of the research and thus gives a picture of how the science changes over a time. I will focus on a few statistics to try to summarize some of the changes that have occurred. In 1987, 4 years after the journal commenced, there were 56 papers published and in 2019, 32 years later, there were 223 papers published in the year. Another change has been the increase in the number of co-authors per paper. For example, in 1987, there were 11 single-author papers, 15 with two authors, 15 with three authors, 8 with four authors, 6 with five authors, and none with more than six authors. In 2019, there were 2 single-author papers, 9 with two authors, 29 with three authors, 42 with four authors, 45 with five authors, and 69 with more than six authors.

Two questions arise from the data. First, does this quadrupling of the number of published papers per year over a period of 32 years signify an equivalent increase in knowledge? Second, how is the increase in the number of authors per paper explained? It is difficult to answer these questions in an objective way. The answers that I propose are speculative. I acknowledge that and am prepared to accept the criticisms that will probably and justifiably be directed my way. I believe that the quality of the published papers has diminished. The increase in the number of authors per paper is a reflection of the lowering in quality.

To expand on the answers, it is acknowledged that science grows by building on the contributions of previous workers. This growth does not mean merely accumulation of more observations but improvements in theories as proposed by Popper. Based on my reading of submitted and published papers, there is now a greater tendency for previous relevant work to be ignored. Many researchers presently rely on keyword searches to identify papers to read. They may miss important relevant research, particularly if their reading is restricted only to recent articles. Also, there is an increasing proportion of papers that are simply accounts of observations with little in the way of theoretical content. Thus, the increase in number of published papers does not equate to a proportional increase in true knowledge.

Based on the proposal of Popper, theories result from imaginative ideas. These usually come from mainly one or sometimes two individuals. The testing may involve mainly one or perhaps two additional experimentalists. What then is the explanation for the increasingly large number of co-authors per paper? Perhaps one answer is that a group of individuals participate in the experimental design and discussion processes. This hardly seems justification for becoming co-authors. Another factor may be the increasing importance of the use of a number of publications in the performance evaluation of scientists. Each co-author can then claim equal merit even if their contribution to the work is minor or even negligible.

I should reiterate that my comments are restricted to only one narrowly focussed journal and not claimed to be applicable to science in general. Have the same trends occurred for other journals? Bauerlein et al. (2010) have drawn attention to the excess of poor-quality research that is presently being carried out and published. It was suggested that the main cause for the steep increase in published articles in scientific journals is the increase in numbers of researchers. Another contributing factor may be the effect of increasing managerialism. Performance evaluation of scientists, often nowadays made by administrators rather than scientists, places emphasis on the number of publications that each produces without sufficient consideration of their quality. This can cause scientists to cut corners by not carrying out due diligence on previous literature and not applying the depth of thought required before hastily submitting their papers in order to enhance their publication record.

References

Bauerlein, M., Gad-el-Hak, M., McKelvey, B., and Trimble, S.W. 2010. We must stop the avalanche of low quality research. The Chronicle of Higher Education, June 13.

MacRitchie, F. 2011. Scientific Research as a Career. CRC Press Taylor & Francis, Boca Raton, Fl.

MacRitchie, F. 2018. The Need for Critical Thinking and the Scientific Method. CRC Press Taylor and Francis, Boca Raton, Fl.

Popper, K.R. 2002. Conjectures and refutations. Routledge, London.

chapter three

Our Failing Education System

A successful democracy depends on its citizens being able to think critically. The alternative is to have a population that is gullible and easily duped. It may form its opinions from sources that feed information that is false. The false information can come from many sources but nowadays, the media, some large organizations, and, sadly, our educational system are particularly influential. In Chapter 2, some concepts that are inherent in the scientific method, which incorporates critical thinking, were described. We should never hold a dogmatic opinion. This allows us to hear alternative points of view and to learn from them. We need to have the capacity to readily admit being wrong and to change or at least modify our views. This requires being able to be detached, i.e. to take the position of a disinterested observer. Scientific thinking is dynamic, never static and is not restricted to scientists. It is the type of thinking that everyone should aim to adopt.

In this chapter, we will begin by looking at two of many current issues that have illustrated deficiencies in critical thinking among large sections of the public. We then move on to examine how educational standards have fallen in several countries, focusing particularly on Australia. Following this, we will discuss what sort of steps might be taken to reverse these trends in educational standards as well as how critical thinking skills may be enhanced.

The Case of Cardinal George Pell

The first issue concerns accusations against a senior official of the Catholic Church, Cardinal George Pell. Cardinal Pell has been the target of a continuous campaign which has culminated in him being accused of crimes against two young boys. The alleged crimes occurred in the 1990s when he was a bishop in Melbourne. One of the boys is now deceased. The accusation of the other boy has led to two trials, an appeal in the Victorian (state) Court of Appeal and finally an appeal to the High Court of Australia. The latter is the final appeal that can be made and its verdict represents the end of the case. The result of the first trial was a hung jury and the second trial gave a verdict of guilty after just over four days of deliberation by the jury. The first appeal against the verdict was dismissed by the Victorian court by a 2/1 decision. The High Court of Australia upheld the appeal

DOI: 10.1201/9781003254065-3

unanimously by a 7/0 decision. Prior to this decision, the Cardinal had spent 104 days in goal.

Although the Cardinal is Australian, the trial and appeal process is of interest internationally because the Cardinal attained a high position as Treasurer of the Vatican. He was, in that capacity, the third person in seniority of the Catholic Church. I do not wish to comment on the legal proceedings apart from summarizing the verdicts as I don't feel qualified. I simply want to examine the public thinking that accompanied the case.

During the time that the case was in process, there seemed to be an overwhelming presumption among a section of the public that the Cardinal was guilty. This may have been partly due to guilt by association. Many crimes, the same as that for which he was accused, were committed by others in the church. The Australian Public Broadcaster (ABC), the charter of which is to be impartial, consistently presented views that assumed his guilt and none of its staff presented the opposite view. The police (in the state of Victoria) trawled for victims (who should have been named as complainants) to come forward although this may not be an appropriate action for police. The basic tenet of the judicial system is that everyone is considered to be innocent until proven guilty. In this case, the final verdict was only reached after the High Court had completed its deliberations. Thus, the comments and opinions expressed in the public arena during the process may have contributed to the prolonged series of trials and appeals. It was also demonstrated by evidence outside the court that it was physically impossible for the crime to have been committed as claimed by the plaintiff. There were many who believed that the Cardinal was innocent but there were only a few who had the courage to speak up.

Anthropogenic Global Warming

The second issue is that of anthropogenic global warming or what is loosely described as climate change. The Earth's climate has always been changing although it requires observations over decades to confirm that the changes in weather are long-term effects and not just short-term aberrations. There is a widely held belief that emissions of "greenhouse gases" are contributing to significant warming of the planet and this could lead to harmful effects on the climate. The emissions mainly result from the burning of fossil fuels (coal, petroleum, natural gas) which have only been occurring since the beginning of the Industrial Revolution. The gases are a mixture in which carbon dioxide, methane, and water vapour are some of the main ones. The theory is that these gases enter the upper atmosphere and add a contribution to the atmospheric layer that controls the escape of heat, causing warming. The effect is generally accepted although it is not known with any certainty how large it is and even whether it can be separated from other variables that affect climate and thus can be measurable. There is a widespread belief among the public that the effect could become large and cause catastrophic effects on the climate. The belief has been promulgated particularly by the Intergovernmental Panel on Climate Change (IPCC) which is an organization funded by the United Nations. This body which is made up of a large number of scientists and bureaucrats has produced regular reports which have had a large impact on public opinion.

The two issues that have been described have no obvious similarity. What has been similar is the shortcomings in the quality of public thinking that has accompanied them. In both areas, there has been a deficiency of critical thinking and a trend towards what has been termed "group think". This can happen when a relatively small number of activists can use emotional language to persuade large numbers of people to adopt specific beliefs. An example of this occurred during the rise of the National Socialist German Workers' (Nazi) party prior to and during World War II. A ministry of propaganda was created which was used for motivating the people to support a nationalistic movement. Ideally, a court trial should be conducted with a minimum of public discussion. In the Western judicial system, an accused person is held to be innocent until found to be guilty beyond reasonable doubt by the court. The case against the Cardinal was conducted during a time when a large number of young people were found to have been abused by members of the church. This fanned a resentment among a section of the public against the church and helped to make the Cardinal a scapegoat. There is a resemblance to the trial of Jesus of Nazareth 2,000 years ago. Although punishment is more humane today, the move to mob rule is still inherent in human nature. This makes it difficult to prevent it having an influence on the judicial process. Critical thinking requires individuals to free themselves from their prejudices and to become detached. This is an idealized state which many of us can only aspire to.

The issue of anthropogenic global warming has been mainly promulgated by the IPCC. The formation of the IPCC owes its origin mainly to the work of Bert Bolin, a Swedish meteorologist in the mid-1980s. Bolin warned of a temperature rise in the coming century caused by human activities, in particular the emission of greenhouse gases. The IPCC was established in 1988 under the auspices of the United Nations. It was created "to provide policymakers with regular scientific assessments on climate change, its implications and potential future risks as well as to put forward adaption and mitigation options". In the interim three decades, the IPCC has provided regular reports which are based on extensive reviews of scientific papers in the field of climate change. The most active of three groups of the IPCC has been the one concerned with "adaption and mitigation options".

The basic premise of the Panel is that anthropogenic global warming has been established and will require actions to be initiated to counter its effects. This is an assumption that many in the scientific community question. However, it is one that must be maintained in order to guarantee continuing funding. As a consequence, the aims have tended to become self-fulfilling. A criticism is that this conflicts with the fundamentals of the scientific method which should be open to self-correction. The large bureaucratic structure of the IPCC which depends on consensus between large numbers of participating scientists is contrary to the scepticism and self-criticism that is embodied in the true scientific method.

In the latest IPCC report, an objective is to reduce the global temperature increase to not more than 1.5°C from that of pre-industrial times. Some obvious questions arise that cast doubt on the viability of this aim. First, is a temperature increase above 1.5°C undesirable? It is not known what the ideal temperature for the planet might be. Increased temperature is often associated with improved health; e.g. death rates in the population are higher in colder weather. There are also benefits to agriculture of higher temperature, e.g. enhanced crop yields. Second, is it reasonable to believe that humans could dial the Earth's temperature to predetermined values? The variables that affect the climate are not fully understood. The models that have been proposed have generally not been successful in accurately explaining observed temperature variation.

What we can conclude from scrutinizing these two issues is that public opinion can become entrenched as a result of persuasive campaigns. This can only happen if a large section of the population do not expose themselves to a sufficiently wide range of viewpoints and are not able to apply critical thinking to an issue. An important aim of education should be to foster critical thinking in younger people. This will make them less susceptible to accepting beliefs before the issues are clarified and fit them to participate constructively in the future of society.

How Have Educational Standards Been Performing?

We will now look at how educational standards have fared in recent times with particular focus on Australia. The Programme for International Student Assessment (PISA) was initiated in 1997 with the purpose of comparing education attainment in countries across the world. Its aim is to provide comparable data to enable countries to improve their education policies and outcomes. Its parent body is the Organization for Economic Co-operation and Development (OECD). Assessment was first performed in 2000 and is repeated every three years. It evaluates educational systems by measuring the performance of 15-year-old students in three areas, mathematics, science, and reading. The tests are administered in the students' first language. The latest results were collated in 2018 and were released in December 2019.

In the period 2000–2018, the performance of the different countries has been variable. For example, in the early stages, Sweden's performance was falling but since then, the trend has been reversed to show a steady improvement. This demonstrates that it is possible for countries to alter their trajectory by introducing the right policy changes. Australia has shown a steady decline from 2000 to 2018 in each of the three areas from initial levels well above the OECD average to levels approaching the average. Over this period, the OECD averages remained fairly stable.

The effects of different variables on performance were also studied. Socio-economic status was shown to be a good predictor of performance. However, a significant number of disadvantaged students were able to score in the top quarter of reading performance in Australia, showing that disadvantage does not always equate to destiny. Another area that was studied was discipline. In Australia, 32% of students (OECD average 26%) reported that, in some classes, teachers had to wait a long time for students to quieten down. In those same classes, students scored lower in reading than students whose teachers did not have to wait. The number of students who reported being bullied at least a few times per month was significantly greater than the average across OECD countries. Examination of all such variables as well as the academic skills provides valuable information for each country to utilize for their future policies.

No attempt will be made to analyse the results for the three areas of learning. This will be left for those who are more competent to do this. Some comments will be given only on the results that emerged outside the three areas. One result for countries in general was that there has been no improvement in performance over the past decade despite a 50% increase in funding. Another important conclusion was that high performance can be achieved even for economically disadvantaged students. The converse is illustrated by the results for Australia. For Australia, the steady decline in performance has paralleled an appreciable increase in the funding for education. Spending on education increased by 79% during the period 2000–2018 while the number of students in the education system increased by only 22%. This means that education spending per student increased by 40% over this time. It shows how misguided were the politicians who criticized governments for not allocating more money for education. Throwing money at an issue is not always the way to improve outcomes. There is an obvious need to look at alternative reasons for failings in the system and how they might be rectified.

Discipline was another factor shown to influence performance. Some students believe that "mucking around" in class shows that they are clever in rebelling against the "evil" authority. They don't understand the importance of discipline and that part of discipline is self-discipline, an important quality in people. In earlier times, teachers were given more authority to punish or reorientate students for their antisocial behaviour. The slackening of standards over recent decades has made it more difficult for teachers to impose order. Lack of order in classrooms is a serious modern-day problem that presents an obstacle to learning. Ideally, cooperation between teachers and parents should be able to resolve issues of bad behaviour. One problem is that some parents are myopic with respect to bad behaviour of their offspring. If bad behaviour of some students is brought to their attention, parents sometimes jump to the defence of their child and take a partisan attitude in the dispute. Both the children and their parents should take responsibility for antisocial behaviour as this impinges on the rights of other students and this should be unacceptable.

Bullying is an undesirable trait that can cause great suffering and can have negative effects on performance. Usually, students who are proficient academically and display other positive qualities such as prowess at sports rarely have a problem with bullies. The targets are those who the bullies perceive as having a weakness. This might be such things as a timid personality, poor academic results, or sporting inability. Paradoxically, bullies usually display cowardly behaviour when confronted by someone who stands up to them.

The Importance of Critical Thinking

Now, let us turn to another striking result that came out of the 2018 assessment. This was that in one test, only one in ten students were able to distinguish between fact and opinion. This brings us back to the two issues that were covered in the first section of the chapter. A common feature of the two issues was the "group think" associated with each, criminal accusations against a Cardinal and the tacit acceptance of significant anthropogenic global warming without a valid justification. What public discussion has shown for both issues is the lack of critical thinking that has been displayed. This is a serious flaw that can lead to acceptance of erroneous beliefs which can cause long-term negative effects on society. If we are to form a society based on justice and wise decision-making, we need to move away from group think towards critical thinking of individuals.

Critical thinking is an essential component of the scientific method. Although many scientific topics are included in the school syllabus, there does not seem to be sufficient emphasis given to the philosophy of the scientific method. It is therefore not well understood. There are many misconceptions about what is involved in the scientific method. In public debate, we regularly hear statements such as "It has been scientifically proven" or "The science has been settled". These statements are false. In science, nothing is ever settled. Science aims to increase our understanding in order to move closer to the truth but it never claims to have attained absolute truth. The way that scientific research proceeds was briefly described in Chapter 2 and this should form a basis in the study of critical thinking.

The inability to distinguish between fact and opinion for students in the PISA programme alerts us to shortcomings in fostering critical thinking in school education. Critical thinking should be an essential topic for inclusion in school syllabi. It should begin by encouraging students to adopt an attitude of humility and respect for the opinions of others. The dangers inherent in dogmatic thinking should be explained. There are many people today who believe that their opinion is the right one and that therefore, anyone with a different opinion must be wrong. This is contrary to what forms the basis of a liberal democratic society. It is an attitude of intolerance that, if allowed to dominate, can lead to totalitarianism in its different forms such as fascism and communism. The failures of such systems need to be illustrated by an exposition of history. If done objectively, it will expose how such systems have destroyed human freedoms and inevitably have harmed large numbers of citizens.

The poor and declining performance of Australian students and the deficiency in critical thinking shown by the PISA results point to the need for rethinking some educational strategies. Let us look at some ways that could be introduced or enhanced to bring about an improvement in critical thinking. Although some schools include instruction on the topic, there seems to be an urgent need to give it greater emphasis.

A course on critical thinking might begin by presenting extracts taken from the media (print media, radio, TV) as points for discussion. Initially, some simple examples of how one can be deceived could be given. Then, some guidance on how to identify faults might be outlined. This could proceed simultaneously with practical exercises for the students to analyse. The extracts should, of course, be relatively easy for them to identify errors in logic or reasoning. Not all the examples should be ones that are defective. Some might be free of faults so that students are also encouraged to recognize truth and fairness. These exercises serve two purposes. First, it gives students practical experience in critical thinking. Second, the analyses and critiques by the students could be passed on to those responsible for the statements, thus making them more careful in their future comments.

We will now look at some typical examples of how faulty thinking can be exposed.

Exercises to Practise Identifying Faulty Thinking

1. Persuasive language: "I am reliably informed that ...".
   Gullible people will be taken in by this sort of statement. Critical thinkers will be sceptical. They will question the source of the "reliable information" and, if forthcoming, will examine its reliability.

2. Exaggeration: "Most people reject this idea …".
   This is a common ploy. A critical thinker will ask, "Has a study been made of a large number of people to justify this conclusion?" or perhaps it is just a ruse to try to convince.
3. The straw man argument: In order to discredit an argument, this ploy covertly replaces the argument by another one and then refutes the substituted argument instead of the original one. This requires alertness to prevent being hoodwinked.
4. Placing labels on opponents: This involves branding someone by names such as racist, sexist, homophobic, far right/far left. The aim is to silence an opponent and can be effective as no one likes to be stuck with a label that may be perceived as negative. The only effective response is to deny the label and question how it can be justified.
5. Casting doubt: This aims to discredit by hinting that someone could have or might have committed a dishonest act although no backup is offered. Some are taken in but a critical thinker will request evidence for the accusation.
6. Erroneous statement: It may be just a number, e.g. 20 million people died of hunger in a country. This will often pass without challenge. It is difficult to counter unless one has done the homework and be in a position to correct the statement. It may even be that the information is not readily available. In that case, the person making the statement needs to be asked to justify the source.
7. The ideological prism: Many people (and this includes media commentators) see problems through a prism that bends the facts to agree with their ideological bias. The only way to counter this is to try to interpret the facts from a neutral position.
   At this stage, it might be useful to stress an important point. A critical thinker must have patience but be quick to realize when it is time to let go. Nothing is gained by pursuing an argument with someone who is not open to conceding a point so it is sometimes best not to persevere.
8. Using absolute terms: It is never the case/it is always the case. Such claims can often be refuted if exceptions are able to be provided.
9. Deflection of an issue: An issue is deflected into another related issue, in a similar way to how a magician deflects attention. A critical thinker needs to be aware of this sort of manoeuver in order to counter it.
10. Ad hominem or playing the man and not the ball. This ploy is used to attack the opponent instead of the opponent's argument. This fault then needs to be pointed out.

Debating

An effective way to stimulate critical thinking is through debating forums. Although debating is an elective topic in some schools, there is probably a need to make it more widely used. In today's world where everyone is bombarded by information, much of which is questionable, it is vital for students to develop skills in debating issues. This helps to fit them to become independent free-thinking citizens and not puppets who allow themselves to be manipulated. A valuable exercise is to reverse the roles of protagonists. What this means is that first, the student argues in favour of a proposition and then, in a later debate, argues against it. This helps the student to see both sides of an argument. In a real-life situation, this person is likely to become more detached and open-minded to other viewpoints and not to adopt dogmatic opinions.

Role of Wonder and Awe in Education

Children are born with a sense of wonder that must not be lost if they are to develop into kind and compassionate people who maintain a curiosity about the world. Many students nowadays become addicted to playing with their smartphones or computers which deprive them of experiencing wonder and awe. Some schoolrooms can be tedious places that can instil boredom. To prevent this, students need to be exposed to exciting events of nature. An obvious one is a star-studded sky on a moonless and cloudless night away from artificial lighting. This may, of course, be difficult to organize for city dwellers and may need an excursion to the countryside. The firmament of vivid whiteness that appears so close is sure to instil a feeling of awe and wonder. The realization that the Milky Way represents the galaxy that we are in impacts the mind. A brief explanation can accentuate the wonder. The reason we see a Milky Way is that the galaxy is shaped like a discus with a long axis whose distance from rim to rim is of the order of one hundred thousand light years. A light year is the distance that light travels (at 186,000 miles per second) in one year. That means that, even if we could get in a spaceship that travelled at the speed of light, it would take 100,000 years to traverse the galaxy from one rim to the opposite one. The opportunity to observe the night sky can be taken to point out some of the constellations and main stars and planets in the galaxy. Another awe-inspiring detail is that there are about one billion stars in our galaxy and that, outside our Milky Way galaxy, there are in the order of a billion other galaxies.

Walks on beaches (for those fortunate to live near the coast) or in bushland can also stimulate wonder. The sights and sounds of birds and the colours and forms of wildflowers and orchids are sure to stimulate the senses. The first time one hears the tinkling sounds of bellbirds (Australian native birds) can leave one deeply impressed. Another exciting activity that some schools have adopted is to plant flower or vegetable gardens. Seeing plants germinate and following their growth can be a fascinating hobby.

Health and Nutrition

The basic facts of good nutrition do not seem to be given the necessary emphasis in school education. The increase in obesity and susceptibility to some diseases such as diabetes are problems that have become more acute in recent times. Deficiencies in diets may be responsible for some of these unhealthy trends. Processed food and aerated drink consumption has increased. Some of these contain high amounts of trans fats and sugars which are known to be detrimental to health. There are three main types of macronutrients in human foods, proteins, carbohydrates, and fats. Also important in health are trace elements and fibre. In the past half-century or so, there has been a movement to reduce fats (and particularly saturated fats that occur in meat and dairy products) in the diet. However, the topic has been controversial and solid arguments have been presented to question whether fats are harmful (Teicholz, 2014). One of the outcomes of reducing dietary fats has been their substitution by carbohydrates and this change has been linked to the observed increases of obesity in the population.

While on the subject of health, it can be valuable to emphasize the destructive effects on people's lives of taking drugs, smoking, and consuming excess alcohol. Unfortunately, graphic illustrations may need to be introduced to shock students into taking these effects seriously.

Bullying

The problem of bullying was emphasized in the latest PISA report. This is a complex and challenging problem that can cause great distress to many people. Bullying is defined as an organized and deliberate misuse of power through verbal, physical, and/or social behaviour that intends to cause physical, social, or psychological harm. This definition is important in order to identify bullies and separate their behaviour from other types of conflict or violence. In schools, bullying undermines values that aim to promote the qualities of respect, trust, and honesty in students. Bullying can cause increased anxiety and unhappiness which often leads to poor attendance records and lower academic performance of victims. Bullies may stop in the short term after a "get tough on bullying" approach but, unless the relationship and social factors are addressed, it is likely to reoccur, take another form, or remain hidden. The long-term aim is to foster a school culture based on supportive relationships, featuring respect, inclusion, belonging, and cooperation.

"Bullying. No Way" is a website for Australian schools that is managed by the Safe and Supportive School Communities Working Group. Its aim is to help schools to create learning environments where every student and school community member is safe, supported, respected, and valued. It is imperative that students who are victims of bullying seek support from such a group. It needs courage to confront a bullying situation. Not doing so can lead to much unhappiness and, in serious cases, to suicide. Students can check out the website to get answers to questions that are commonly asked. If this is not sufficient, mailing and e-mail addresses can be found on the site.

Those who bully also need help. Otherwise, they may come to see bullying as normal and acceptable. They need to learn more appropriate ways to act so as to avoid developing more serious antisocial behaviour. Bystanders who witness bullying also need counselling on how they should behave. If they do nothing, it might be seen by the bully as a sign of approval. They need to receive training on how they should intervene safely.

For those who are involved in the subject of bullying, they may be interested in a book entitled "Auggie & Me. Three wonder stories" authored by Palacio (2015). The story has also been made into a film. It describes the experiences of a young boy, August Pullman. Auggie, as he was called, was born with a facial deformity. How he coped with the affliction and how his schoolmates responded to it make inspirational reading.

The Role of Physical Activity in Learning

There is a saying that if you find yourself getting deeper in a hole, then you should stop digging. The steady decline in the education performance of Australian students shown by the PISA programme suggests that the system may need some changes. Declining standards can be reversed as shown by the example of Sweden. It is probably naïve to think that a single change could turn things around. However, policies based on sound research are worth considering.

John Medina is a development molecular biologist who has some interesting views on how the brain develops based on his research. He points to the fact that physical activity has been crucial to the evolution of the human brain. The subjection of students to spend long sedentary periods in classrooms is contrary to the best conditions for learning. He recommends that the design of classrooms and the methods of instruction place more emphasis on exercise. Research has shown that learning is enhanced when class teaching is interspersed by periods of physical activity. Medina has written several books (e.g. Medina, 2018) and educational tools that can be utilized in schools.

The Importance of Personal Hygiene

A positive outcome of the COVID-19 pandemic has been the greater attention being focused on personal hygiene habits in avoidance of disease. There has been more awareness in the general public about how disease can be transmitted through the community. There are two main ways that infections can be spread. One is by inhaling germs by close proximity to people who are sneezing, coughing, or just breathing without taking precautions to prevent airborne transmission. The other is by touching contaminated surfaces and not washing hands thoroughly (with soap for at least 20 seconds) before touching the face or before consuming food. Awareness of the need to avoid these two sources of contamination has increased in the general public. The avoidance of close contact with others and regular hand washing can reduce the risk of infection.

References

Medina, J. 2018. Brain rules for aging well. Scribe Publications, Carlton North.

Palacio, R.J. 2015. Auggie & Me: Three Wonder Stories. Knopf Books for Young Readers, New York.

Teicholz, N. 2014. The big fat surprise. Simon & Schuster, New York, NY.

chapter four

How Political Debate Should Work in a Democracy

In the previous chapter, there was a discussion on ways to enhance the educational standards of a nation. One topic that was omitted but which is crucial to success in a democracy is political education. Public ignorance in this area is striking and is inappropriate for a successful society. The deficiency of knowledge and good judgement in political matters impacts the population and ultimately on society. This deficiency especially becomes apparent around the time of the election of a government. The community separates into various groups. Members of one group devote a good deal of time to evaluating the current government's record and the policies they propose to follow if elected. They also evaluate the criticisms brought up by the opposing parties and the policies they propose to initiate if they are elected. Outside this group, whose members could be classified as critical thinkers, there is a spectrum of citizens in the electorate who contribute to the result of the election and thus are at least in part responsible for future policies.

The spectrum of voters can be roughly separated by different characteristics, some of which overlap. Some people give little or no thought to the election and, if voting is compulsory, turn up at the polling station to avoid having to pay a fine. If they vote, they may just fill in numbers (a donkey vote). They are not interested and don't care who wins. In a system where voting is optional, I have heard some say such things as that they don't trust politicians so they don't vote or that it won't make any difference how they vote. Cynicism and failure to take responsibility are examples of actions that threaten a democratic system. There are those whose families have always voted for a certain party so that is the choice that they blindly follow. Some people don't feel that they have time to devote to consideration of issues and may be guided by hearing a slogan or the opinion of a relative. Others may be influenced by hearing views expressed on a media outlet. The media outlet that they choose to heed is quite likely to be partisan. Many people who evaluate the different parties cast their vote for whom they believe will further their own personal interest and will give little consideration to what is best for the public good.

If this behaviour is how an appreciable section of the population approaches political decision-making, it seems obvious that it is not the

DOI: 10.1201/9781003254065-4

ideal way. A successful democracy needs a public who is well-informed and takes responsibility for its formation. There are people who take a pride in being apolitical. Education in political maturity should begin in the schools. If done appropriately, it might phase out much of the apathy and cynicism that exists and replace it by acceptance of the responsibility that defines a valuable citizen.

Many different political parties exist in democratic countries but, usually, the main political movements can be divided into two, one denoted as the left and one denoted as the right. To put it in simple terms, the aim of the right is to grow the pie (i.e. the economy) and the aim of the left is to share the pie more equitably. This of course is a gross oversimplification. The policies of each side will be amplified later in the chapter but the division will suffice to enable us to proceed. Each side of politics has its positive virtues.

A successful democracy depends on a respectful battle of arguments between the two sides. Where it can go wrong is when the system becomes unbalanced. This can happen if it moves too far to the left or too far to the right. History has shown how movements to extreme positions can have disastrous effects. The extreme right movement of Nazism in the early part of the 20th century and the rise of the extreme left of Communism around the same time both resulted in totalitarian regimes with accompanying misery and millions of deaths. It has been suggested that the political spectrum can be best described by a horseshoe rather than a linear left-right continuum with Stalin on the extreme left and Hitler on the extreme right. Although there has not been unanimous agreement on this, it does have some credence in that both ideologies approach a common trait of totalitarianism.

How then should students be instructed on how to develop their political awareness? A problem in today's schools is that a large proportion of teachers have adopted a political standpoint and have tried, with some success, to impose their views onto students (Marcus, 2016; Hopkins, 2017), an activity that is called brainwashing. If we are to educate students politically, it is imperative that the instruction be carried out in a non-partisan manner. This means that the students must be given complete freedom to form their own views and not be manipulated by the teachers.

A promising mode of instruction has been used by Kate Habgood (Habgood, 2016) to develop student political awareness. She introduces the subject by asking students to draw the horseshoe that was mentioned earlier. The good characteristics of left and right ideologies are first explained. The horseshoe is then used as a template to record the students' political views on different issues. The important point in this exercise is that students are encouraged to form their own opinions free of any outside coercion. These opinions are flexible and can change over time.

Let us look at some of the issues that are currently the subject of debate, with the emphasis on the Australian scene and contrast the positions adopted by each side of politics. For each issue, the positions favoured by the left and that favoured by the right are summarized. In place of left and right, we could substitute the terms progressive and conservative. The positions given in many cases are the extremes and are not intended to represent the views of all individual members of the two sides.

Left (Progressive)		Right (Conservative)
	Issue: Immigration	
Open borders		Strict immigration controls
	Issue: Government	
World government		National governance
	Issue: Climate change	
Anthropogenic global warming		Scepticism about human influence
	Issue: Marriage	
Marriage equality		Traditional marriage
	Issue: Renewable energy	
Subsidise renewables		Renewables to compete in market
	Issue: Coal	
Phase-out coal		Maintain coal in energy mix
	Issue: Nuclear energy	
Ban nuclear		Develop nuclear energy
	Issue: Government	
Big government		Small government
	Issue: Bureaucracy	
Increase to administer programmes		Decrease to foster independence
	Issue: Education	
Preferential treatment for public schools		Support for private schools
	Issue: Divorce	
Facilitate divorce		Strict divorce laws
	Issue: National history	
Evils of colonialism		Pride in heritage
	Issue: Genders	
Multiple genders		Traditional genders
	Issue: Abortion	
Pro-choice		Pro-life
	Issue: Euthanasia	
More laxity for voluntary suicide		Strict controls
	Issue: Unions	
Greater power		Curb union power

Left (Progressive)		Right (Conservative)
	Issue: Criminal justice	
More leniency		Capital punishment an option
	Issue: Religion	
Not taught in schools		Included in school instruction
	Issue: Contraceptives in schools	
Greater availability		Greater parental control
	Issue: Israel	
Villain of Middle-East		Lone democracy in region

These are 20 issues that can be used as a basis for students to think about. Most are not black and white. If a horseshoe is adopted as the template to record their positions, they are free to choose a spot intermediate between the two ends. They may settle on a position that is closer to one end (left or right). The role of the teacher is to offer guidance in a non-partisan manner. As the course proceeds, the students are free to shift their position on issues as they carry out more research and give the issues more thought.

There are many tests available on the internet for students to check their political position at a given time. Two sites that may be useful are:

- Political Left/Right Test – Individual Differences Research Labs (IDRlabs.com). One of the choices at this site is Political Test. This has been crafted by experts and provides 36 questions on which to make an assessment.
- The political compass test (www.politicalcompass.org).

Indicators of Moves Towards Extremism

There are several criteria that can warn when there is danger of a move towards extremism in a democracy. We will look at four of these described in the following sections.

Refusal to Accept the Result of a Democratic Election

In a stable democracy, the loser in an election accepts the will of the people with grace and promises to work for the success of the nation. In recent times, there have been examples where this custom has not been so apparent. The 2016 U.S. election produced a result that was not anticipated by many citizens and especially by a section of the media. Similarly, the 2019 Australian federal election gave an outcome that surprised a good many people. In both cases, a party of the right was successful. There have been attempts to rationalize these results. It has been suggested that the public

could not have understood the policies or that there hadn't been sufficient time to explain them. This demonstrates a dogmatic and intolerant
attitude. Some have said that the results must mean that those who voted
for the successful party are too dumb to understand the issues. This is a
hypothesis that could be tested, at least in theory. We saw in Chapter 2 that
a valid scientific hypothesis is one that is testable and refutable. Thus, we
could carry out a survey of the two groups who voted oppositely and measure their intelligence by an IQ test. A sufficiently large sample would be
taken to satisfy statistical requirements for reliability. We cannot say with
certainty what that would show but it would be remarkable if a significant
difference in average IQ would be found between the two groups.

Attempts to Silence Opposition

A reliable indicator that there is a movement towards an extreme position
is the tendency for one group to try to silence those who disagree with its
views. Currently, we are seeing attempts to shut down debate and prevent
the views of opponents from being heard. This is being done by shouting
down or using intimidation and violence.

Lack of Humour

Humour is a measure of a balanced mind. Groups where we see it lacking
can be a useful indicator that there is a move towards extremism. People
who are moving towards totalitarianism can be sarcastic and scornful but
are lacking in a sense of humour.

Distortion of Facts

Another indicator for those who move towards extremism is exposed by
their distortion of facts. This may not necessarily be a blatant attempt to
lie although that tactic has been deliberately used by propagandists to further their cause. It is frequently the result of having an ideological prism
in the mind. This causes facts that arrive to be refracted by the mind of
the recipient so as to accord with their ideological beliefs. This trait can be
reinforced by brainwashing or groupthink.

How Scientific Thinking Might Contribute to Political Debate

In some political debate, there unfortunately seems to be an increasing
animosity between participants. Of course, the debate should be robust
but it should not be disrespectful. On occasions, it can degenerate into

simple point scoring and scorn towards opponents. How can the quality of political debate be improved? As we have seen, democracies can work well when there is a contest of ideas from both left and right sides of the spectrum. It can fail when the balance is shifted too far to one side, i.e. towards extreme ideology. What may help to prevent this tendency is to apply more of the thinking that is inherent in science. What are some of the attributes that have enabled science to produce the advances that we have seen in many areas such as agriculture, medicine, transport, and communication to name just a few of the many areas.

In Chapter 2, the scientific method was described based on the philosophy of Karl Popper. In principle, the scientific method is simple. A challenging problem is confronted and a hypothesis is proposed to try to explain it. An essential criterion for a scientific hypothesis is that it must be refutable; i.e. it must be falsifiable. This distinguishes it from pseudoscience. Examples of pseudoscience are the predictions made by astrologers. These are similar to those of fortune tellers. The predictions of fortune tellers are made so vague and general that they cannot be refuted. Irrefutability makes a theory non-scientific. Science thus proceeds by proposing hypotheses, testing them, and rejecting those that are refuted. Those that are not refuted, even by severe tests, are said to be corroborated but with the proviso that they may be refuted at a later time. Scientific knowledge therefore advances through a trial and error procedure. The rationale is that there is an approach towards truth without scientists ever claiming to have arrived at absolute truth.

In an ideal scientific debate, the protagonists free themselves from bias and state their arguments, each accepting that they don't possess the absolute truth and are willing to listen attentively to the opponent's argument. Of course, this is an idealized situation but is one that scientists (and everyone) should aim to aspire to. A criticism that might be made of scientists is that they don't hold convictions like those who, for example, have faith. This follows from the admission of scientists that they do not profess to know the absolute truth and that they always remain open to doubt. To some extent this is true. Those who have faith are unshakeable in their belief whereas scientists (true scientists) never claim to be certain. Scientists, nevertheless, can have strong opinions but they are based on reason, not on faith and are ready to change them if new results or new arguments emerge.

Is it better to have tentative convictions or unshakeable ones? What if the unshakeable ones happen to be wrong? You are then trapped with nowhere to go. Of course, you can ignore the new results or arguments and adhere to your beliefs. Some people, even some who call themselves scientists, do this. For example, a scientist may propose a hypothesis to explain a certain phenomenon. When experiments are done to test it, they show that some facts do not agree with the hypothesis. However, some "scientists" are reluctant to give up their hypothesis. After all, it was based on their original creative thought so they feel a certain ownership of it. Perhaps no one else in the world had thought of the idea, so they may find it hard to relinquish. Some results may not agree with the hypothesis but these can be swept under the carpet. Out of sight is out of mind. This is a dangerous thing to do. These "scientists" may never escape from their prison. In contrast, scientists who readily admit that the hypothesis was wrong and reject it, free themselves from a false belief. They can then learn from their mistakes and use this knowledge to formulate an alternative (and usually better) hypothesis.

Applying the Scientific Method to Issues that Involve Science

In the list of issues that are debated in the public domain, some require the application of science. Climate change is an obvious one. It therefore seems remarkable that opinions divide on a political basis. There is general acceptance on scientific considerations that gases produced by burning fossil fuels in industrial processes could contribute to a warming effect. This is caused by these gases concentrating in the atmosphere and participating in a sort of blanket that reduces the escape of heat. The gases, referred to as greenhouse gases, are a mixture of which water vapour, carbon dioxide, methane, nitrous oxide, and ozone are the main ones. The point of contention is how large their effects are on the Earth's temperature. The political left largely supports the view that there is a significant human contribution to a changing climate and that actions are needed to counter this trend. Those on the right are more sceptical. Why should there be a division on political grounds to an issue that is primarily scientific? We are going to look into this apparent paradox. Before doing so, we will revisit some of the basic characteristics of the scientific method.

Scientific enquiry has existed for a long time. Its progress received special impetus during the Enlightenment period, notably between the 18th and 20th centuries. It was characterized by enquiring minds, measurements of observations, and the rejection of dogma in favour of reasoning. Thomas Kuhn (Kuhn, 1962) proposed that science advances, not by a regular progression of new knowledge but by periodic transformations that have been called paradigm shifts. One of these paradigm shifts has been in relation to the philosophy of the scientific method which has been enunciated by Karl Popper (Popper, 2002) in the 20th century. Popper exposed how induction was not an effective method for furthering true knowledge. In simple terms, the inductive method records observations and then attempts to arrive at a general encompassing explanation of them.

Popper showed that the true method for advancing knowledge was that of hypothesis-deduction. This method involves a creative step of making imaginative conjectures followed by experiments designed to severely test them. Instead of looking for confirmations, the crux of the hypothetico-deductive method is to try to falsify the conjecture (or hypothesis).

This is a crucial point of departure of the two methods. By replacing attempts to find a confirmation by attempts to find a falsification, this leads to a more open-minded approach. In the inductive method, the observations that are made tend to be arbitrary. In contrast, in the hypothetico-deductive method, observations are not arbitrary but are deliberately designed. They are thus observations that would never have been conceived if it were not for the initial hypothesis. Thus, unlike induction, the scope of observations has no limits and allows the investigator's mind to soar into previously unimagined territory. This difference in the two methods is captured by a quotation attributed to Albert Einstein. Einstein said, "Imagination is more important than knowledge. For knowledge is limited to all we know and understand while imagination embraces the whole world and all there ever will be to know and understand". Induction can be thought of as encompassing knowledge whereas hypothesis-deduction embraces the additional concept of imaginative thinking.

Anthropogenic Global Warming

In recent times, it has been observed that the Earth's temperature is increasing and a view has formed that this has been partly due to human activities. The main activity has been the burning of fossil fuels (mainly coal) that has been occurring since the start of the Industrial Revolution in the early 18th century. This human-originating effect (termed anthropogenic global warming) has caused concern that it may lead to dangerous consequences for the planet. The topic has produced considerable media comment and resulted in significant international political action aimed at countering the assumed effects. A history of the key events that have occurred in regard to climate change for the period 1712–2013 has been succinctly summarized by Richard Black (Black, 2013) for the period 1712–2013.

In 1988, the Intergovernmental Panel on Climate Change (IPCC) was formed under the auspices of the United Nations to collate and assess evidence on climate change. Its first Assessment Report was produced in 1990 and there have been regular reports up to the 6th produced in 2018. The First Assessment Report concluded, "temperatures have risen by 0.3–0.6C over the last century, that humanity's emissions are adding to the atmosphere's natural complement of greenhouse gases, and that the addition would be expected to result in warming". The 3rd IPCC Assessment

Report found "new and stronger evidence that humanity's emissions of greenhouse gases are the main cause of the warming seen in the second half of the 20th century". During the period 1958–2008, the concentrations of carbon dioxide (CO_2) measured at Mauna Loa, Hawaii have risen from 315 ppm to 380 ppm. The latest figures give a value of >400 ppm. It was also noted that the human population of the world has steadily increased from 1 billion in 1800 to 7.6 billion in 2018.

It is not the intention here to try to analyse the relevant data that have been recorded. The aim will be to examine the issue from the perspective of the philosophy of the scientific method and to point out where there have been some failings. As mentioned earlier, there has been a dispute between those who support the view that there is a significant (measurable) anthropogenic contribution to global warming and those who are not convinced that this poses a serious threat. A difference of opinion is perfectly acceptable and, in fact, desirable in science. In an ideal scientific debate, those on each side put forward their arguments, accepting that they may not possess the truth and are willing to listen respectfully and consider opposing arguments.

Let us look critically at the thinking that has been used on both sides. First, there is general agreement based on science that human activities such as burning of fossil fuels produces greenhouse gases and their concentration in the atmosphere should make a contribution to a warming effect. What is not known with any certainty is if this is a large effect which could have disastrous consequences for the Earth's climate, whether it is so small as to have negligible effect or if it is somewhere between the two extremes.

Suppose, we apply the hypothetico-deductive method and postulate that the Earth's warming is due in large part to industrial emissions of greenhouse gases. Can we then devise an experiment that, in principle, could refute the hypothesis? If we can and the experiment refutes the hypothesis, then it is rejected. If the experiment fails to result in a refutation, then the hypothesis is corroborated although we can never claim that it is proven beyond doubt. The experiment would go like this. All industrial processes would be shut down for a period, say three decades; i.e. no fossil fuels would be used. The concentration of CO_2 in the atmosphere would be monitored during this time as well as the Earth's temperature. If at the end of the period, the CO_2 concentration was found not to have risen and the Earth's temperature was still continuing to rise, it would amount to a refutation and the hypothesis would be rejected. If the CO_2 concentration and the Earth's temperature both did not increase, the hypothesis would be corroborated.

Obviously, such an experiment is completely impractical and not meant to be taken as a serious suggestion. It is simply described to illustrate the type of experiment needed to test the theory of anthropogenic global warming and the difficulty of reliably testing such a theory; i.e. the theory that human activities make a significant contribution to a global temperature increase. There are, however, other variables that would need to be considered. Industrial emissions are only part of the anthropogenic contribution of greenhouse gases although perhaps the main ones. There are, in addition, contributions from agriculture and transport. Furthermore, the experiment envisaged would only be definitive provided anthropogenic effects and global temperatures were the only two variables to be considered. Unfortunately, it is not so simple. There are other variables that are superimposed. Some of these variables may not even be known and, if they are known, their effects may not be properly understood. For example, the sun's variability, volcanic activity, and changes in ocean currents are variables whose influences may not be able to be quantitatively allowed for with anything approaching certainty.

In spite of these difficulties, the experiment suggested may be the most perfect one that can be visualized to provide, at least in principle, a rigorous test of the theory of anthropogenic global warming. As we have seen, observations based on the inductive method are invalid for acquiring true knowledge. We can point to observations that are consistent with the theory. For example, it has been reported that glaciers in the Himalayas have been melting rapidly in recent times (Carrington, 2019). There have also been reports that sea levels have been rising (Church and White, 2011). Both these observations are consistent with warming and it has been pointed out that their timing coincide with an increase in industrial emissions. However, an important point to recognize is that correlation does not necessarily signify a cause-effect relationship. Furthermore, the Earth Science Division of NASA Space Flight Center has reported that the Antarctic Sea ice has been increasing in size over recent times, a result which is opposite to what would be expected if warming were occurring. Thus, if we look for confirmations that the Earth is warming, we can find them. If we look for confirmations that the Earth is not warming, we can also find them. This shows the inadequacy of the inductive method.

The Hypothetico-Deductive Scientific Method

One of the most classical examples of the hypothetico-deductive method was its application to testing the theory of general relativity, conceived by Albert Einstein. One of the predictions of this theory was that light should be deflected by gravity and the theory could predict the direction and magnitude of the deflection. In order to test the hypothesis, an experiment was suggested in which the light from distant stars would be measured when it passed a heavy body, the sun. This experiment could not be performed during day time as the stars cannot be seen due to the sun's brightness.

At night, the stars can be seen but the sun is not present. It was therefore decided to wait for a total eclipse of the sun. When this occurred, the positions of stars in the vicinity of the sun could be seen and their positions photographed. Comparison of their positions in the night sky would show whether the light was deflected and the direction and magnitude of the deflection compared to what were predicted from the theory of relativity. The results in fact did confirm what the theory predicted. The hypothesis complied with the criterion that it needed to be refutable. Even if it had been found that the light was deflected but the direction and magnitude of the deflection had been different to what was predicted, the hypothesis would have been refuted.

The Merit of a Hypothesis

Returning to the discussion of gradation in the merits of hypotheses, it is evident that predicting that light is deflected by gravity according to the theory of relativity can be ranked highly. Even if the hypothesis had been shown to be wrong, it is still scientifically a good one because it is testable and refutable. In contrast, the predictions of astrologers and fortune tellers are not refutable. If it is predicted that something good will happen to you tomorrow, this is not refutable. Waking up in the morning could be taken to be a good thing and it is almost certain to happen. The difference in the predictions of the theory of relativity and those of the fortune teller is related to probability. A good scientific hypothesis such as that light would be deflected by gravity and the direction and magnitude of the deflection estimated was not expected so it had a low probability. Predictions of fortune tellers are made so general that they are almost certain to happen so they have a high probability.

How a Belief Can Be Imposed on an Uninformed Public

There is presently a forceful movement to persuade the world that science shows that significant anthropogenic global warming has been proven. As a consequence, it has been accepted by a large proportion of the public, including the media. This widely held belief is not properly based on science but is passed on and accepted by bureaucrats and politicians who then determine policies based on the belief. The policies can have enormous effects, quite often detrimental. It may be that it will be shown that human activities are having large effects on climate. However, it is not true that it has been established based on science. It is being uncritically accepted and promulgated by people who have influence but are not scientists or who do not have an understanding of science.

How Can We Determine if Anthropogenic Global Warming Is Significant?

There is a need to address the question of how to reliably determine whether human activities are having a significant effect on climate. Until now, this has not been done even though enormous economic costs have been imposed based on the assumption that it has. What if it is found that the contributions of human activities are negligible and all the expenses that are being assigned to the assumed problem are unnecessary and therefore wasted? Surely then it is worth trying to devise a method that could resolve the question based on sound science. The inductive method often based on correlations can be ruled out. The experiment in which human activities that contribute to greenhouse gases are terminated, although acceptable as a scientific hypothesis, is also ruled out because of impracticality. Therefore, we need to come up with other experimental tests. These must be based on the hypothetico-deductive method to have validity. It is not always necessary to plan and implement an experiment to test a hypothesis. It is sometimes possible to do this by examining events that have occurred in the past.

Possible Approaches to Finding an Answer

A Hiatus in Global Warming

We ruled out the experiment to terminate human activities that produce greenhouse gases on the basis of impracticality. What about if we reverse the experiment and terminate global warming to see whether CO_2 (a marker for greenhouse emissions) stops increasing. How can we do that? We can't. However, Mother Nature might assist in such an experiment. There is a record of climate change throughout history involving long-term fluctuations in Earth's temperature. In recent times, there have been periods when the temperature appeared to stay fairly constant, referred to as a hiatus. It was claimed that there was a hiatus in the early part of the 21st century although it has been argued that, when other variables are taken into account, there has been no change in the long-term warming trend. The debate is not easy to resolve as other variables are superimposed. It usually needs observations over long periods of the order of at least 30 years to confirm a definitive change in climate.

The sun goes through changes in energy output, some occurring over regular 11-year periods. At times, it becomes quieter for longer periods. In Roman times, there was a solar minimum from 450 to 700 AD. More recently, there have been other periods, the Maunder minimum (1645–1715) and the Dalton minimum (1790–1830). The last Grand Solar Minimum (1650–1715) coincided with a period called the Little Ice Age.

The science that governs fluctuations in the sun's emission of energy is not well understood. Other variables that can affect climate such as volcanic activity are also not predictable. It therefore seems reasonable to proceed with caution before introducing policies that have large impacts on humanity without having a solid scientific justification.

The Medieval Warm Period

It has been shown that there was a period between the 9th and 15th centuries in which the rate of warming and the temperatures reached in the Northern Hemisphere approximated those that have been observed in recent decades. If it were shown that similar effects occurred in the Southern Hemisphere during these times, it would indicate that it was a global phenomenon. If this were the case, then warming in recent decades could be seen as natural climate variations rather than driven by industrial emissions. Some experimental measurements to test the idea have been suggested by Asten (2010) using fossil-shell, cave-deposit, and tree-ring records from Tropical to Antarctic Australia. In fact, some experimental confirmations have been reported in South America (Neukom et al., 2011; Bracco et al., 2011).

In the research by Neukom et al., a reconstruction of the mean temperature history of a period between 900 and 1995 was made for a region called Southern South America. This showed that there was a warm period from about 900 or earlier to the mid-14th century. During this period, a peak in temperature was found between 1079 AD and 1089 AD that was estimated to be about 0.17°C above the peak reached in the current warm period. The report by Bracco et al. was based on a study of phytoliths (microscopic plant fossils) in Uruguay. It was concluded that Uruguay's climate from 750 AD to 1350 AD was warmer than the current climate.

Global Warming-Carbon Dioxide Correlation

There are indirect ways of measuring fluctuations in CO_2 concentrations and temperatures in the past, over periods of thousands of years. Therefore, we could, in principle, use this data to test the theory. The data that have been collected indicate that there has been a correlation between CO_2 concentration and temperature over past cycles. However, the data have shown that increases of temperature have usually preceded increases in CO_2 concentrations, sometimes by very long periods, opposite to what is predicted. The topic has been controversial and it has been difficult to arrive at a firm conclusion. One of the problems is that the levels of both CO_2 concentration and temperature can vary considerably from place to place and from time to time, making it difficult to obtain reliable averages.

References

Asten, M. 2010. CSIRO should establish if there was medieval warming down-under. The Australian, May 13. www.theaustralian.com.au/opinion/csiro-should-establish-if-there-was-medieval-warming-down-under/news-story/55aafb98e482713e41c1f16b5e7eaba3

Black, R. 2013. A brief history of climate change. BBC News, 20 September.

Bracco, R., del Puerto, I., Inda, H., Paniero, D., Castineira, C. and Garcia-Rodriguez, F. 2011. The relationship between emergence of mound builders in SE. Uruguay and climate change inferred from opal phytolith records. Quaternary International 245: 62–63.

Carrington, D. 2019. Himalayan glacier melting doubled since 2000. The Guardian, June 20.

Church, J.A. and White, N.J. 2011. Sea level rise from the late 19th to the early 21st century. Surveys in Geophysics 32: 565–602.

Habgood, K. 2016. Political literacy in Australian schools. Breaking Out: Journal of Schools, Community and Social Justice 1(1): 24–26.

Hopkins, K. 2017. Schools are supposed to teach kids HOW to think for themselves not WHAT to think. SO why are so many liberal teachers bullying and brainwashing children with their own intolerant view. Daily Mail, February 6.

Kuhn, T. 1962. The structure of scientific revolutions, 3rd edition. The University of Chicago Press, London.

Marcus, C. 2016. Why are Australians becoming dumb and dumber? Ask the teachers. Daily Telegraph Melbourne, December 20.

Neukom, R., Luterbacher, J., Villalba, R., et al. 2011. Multiproxy summer and winter surface air temperature field reconstruction for southern South America covering the past centuries. Climate Dynamics 37: 35–51.

Popper, K.R. 2002. Conjectures and refutations. Routledge, London.

chapter five

Fairness and Equality

Chapter 1 touched on the subject of resilience. This is a quality that should be nurtured if our aim is to create a better society. In Chapter 1, I reflected on my own experiences as a child growing up. The conditions were such that they required a certain amount of fortitude to contend with the hardships that I faced. These hardships were far from formidable and probably minor in comparison to those faced by most other children in the world at that time and certainly previous times. I have sympathy for young people of today who are not presented with the same degree of challenge. Many children these days are mollycoddled and protected from having to face harsh situations. This acts as a barrier to development of their resilience.

The acquisition of resilience, in addition to fitting young people with the ability to cope with difficult issues that they are certain to face in later life, helps to develop other qualities. It helps them to accept their lot. It helps them to recognize and accept that there are some who are worse off and others who are better off than them. Persons who have built up resilience are likely to feel compassion for those who are worse off. They are also less likely to feel envy for those who are better off. Instead, they may feel admiration and it may stimulate them to strive for improvement of their own conditions. This is all part of forming a positive attitude to life. Such people are more likely to make useful contributions to society. In contrast, those who have not developed resilience or have not had the opportunity to do so are more likely to adopt negative views. They will be more prone to become jealous of those who appear to have more in terms of wealth, comfortable lifestyles, and opportunities.

Equality of opportunity is something that most people support. However, it may not always occur. When it doesn't, the most positive reaction often may be to accept that justice is not always served and to make the best of our lot in the circumstances. Equality of outcome is an entirely different thing. The only way this can be achieved is by force; in other words by imposition of tyrannical rule.

Understanding Fairness and Equality

How do we understand the concept of fairness? We will take a simple example as a basis for discussion. After the Australian federal election in 2013, the elected government proposed a budget. It was based on the

DOI: 10.1201/9781003254065-5

proposition that there had to be measures taken to restore the economy to a path for prosperity. This involved asking all sections of the nation to make small sacrifices. Some of these measures unfortunately had not been foreshadowed prior to the election. As one example, pensioners who had previously been eligible for free medical treatment would be asked to contribute $7 for each doctor consultation. After 12 consultations in the financial year, they would then become exempt from further payments. The proposed change was taken advantage of by those who opposed the government, which included the opposition in parliament, sections of the media and the head of the Australian medical association. One of the arguments used was to brand the proposed budget as "unfair". The word "unfair" was used with great success to convince a large section of the public and to force the government to back down from the proposal as well as others that had been foreshadowed in the proposed budget. The net effect was to remove many of the reforms that, if they had been implemented, would have benefited the nation without placing an excessive burden on any section of the community.

This brings us to the question "What is meant by unfair?" On one side of the debate (the opposition), the argument was that poorer people (pensioners) would be unfairly targeted by asking them to make a contribution. The government argument, although not vigorously put, was that fairness should mean that responsibility be shared by everyone. The contribution of the poorer people in the community (pensioners) was relatively small in proportion to their economic capacity. However, the fact that they were the poorest acted as a political lever for the opposition to attack the government's proposal. The argument, taken up by the opposition parties, the leader of the medical association (whose political views aligned with the opposition) and major proportions of the media and public was sufficiently powerful to cause the government to back down.

What lessons can we learn from this example? First, interpretation of fairness is often subjective and difficult to measure in an obvious way, at least in the view of a large section of the community. Individuals will differ in their opinions of what is fair and what is unfair. It may come down to one's perception of how a policy may affect them personally. Second, this disagreement on what constitutes fairness can be used politically to block legislation that could be beneficial for the economy. Third, a government needs to be wary of proposing measures that may have positive effects for the nation but can be quashed by opposing arguments based on the subjective concept of fairness.

I was recently posed the question "What have been the main changes in society that you have noticed during your lifetime?" Without having much time to ponder the question, I gave a reply along the following lines. When I was a child, living in a rural area, there were a few families that struggled to survive. The community was a closely knit one and the plight of these families was known. Meetings of townspeople were held in which ideas about how these families might be helped were discussed and actions put in place to do this. Today, when we see someone who is a millimetre taller, we think "This is not fair. Why should someone be taller than me?"

This is obviously a gross exaggeration but what I am trying to convey is that I have perceived a change from a caring community to one more preoccupied with relative status. This of course is a sweeping generalization. There are still plentiful examples of kindness occurring. My recollection is that there was a genuine concern for those who were worse off. For those who were better off, we were pleased for them and it encouraged us to work harder to become more like them. In short, I had the feeling that the earlier times were characterized by more compassion and less envy.

How Should One Deal with Inequality?

In a free society, there will always be some who are worse off and some who are better off. How should we regard this situation? One way is to protest against this apparent inequality. We could strive to bring about a lowering of standard of those who are better off. How did these people become better off than others? Perhaps they benefited from an inheritance. Perhaps they were lucky by winning a lottery. They may have made more intelligent choices in life, they may have studied to develop special skills or may have just worked harder. There are myriad possible reasons. Is it better to have a society in which there is this apparent inequality between its members or one in which everyone has similar economic status? If we opt for the latter, how do we achieve it? Aspirations of people vary. Some aspire to attaining riches whereas others are content with their present financial status and have no driving ambition to improve it.

Another consideration to be taken into account is the contribution that each individual makes to the country's prosperity. Australia's richest person, Gina Rinehart, head of Hancock Prospecting, a company whose main activity is iron ore production, paid $860 million in taxes in the 2017–2018 financial year (Thompson, 2018). This made her Australia's highest taxpayer. She paid almost $5 billion in taxes over the past 8 years. Mrs. Rinehart primarily invests in Australia while providing employment and opportunities for thousands of Australians. The creation of wealth requires entrepreneurship and preparedness for risk-taking. Should we suppress aspiration and the fruits of innovation in order to make everyone more equal?

Economic wealth is one way to measure equality but there are others. For example, individuals can vary in the knowledge they possess. Would it be fairer to distribute knowledge equally to everyone? We could take the special knowledge of top surgeons and spread it so as to increase the knowledge of everyone in the area of surgery by a small amount. Similarly, the engineers who design airliners and the pilots who fly the planes have knowledge that could be shared so that these specialists would have less but everyone else would gain a snippet of knowledge. That is the case because total knowledge, just like total wealth has a finite value at a given time. We could extend this reasoning to other areas. Is it fair that some people have greater knowledge or greater proficiency in art, music, or literature? Some have greater knowledge of history and science. How do we deal with this inequality? It should be obvious that this line of argument is absurd. The formation of a society in which its members are free to achieve their aspirations has to be superior to one which applies force to make everyone equal. Meritocracy must obviously be a better goal than mediocrity.

How Do We Understand Fairness?

In my youth, I was keen on sport. One sport I played was cricket. In Australia, I didn't play very much. I found having to spend long times in the hot sun not very appealing. When I moved to England and spent three summers there, I found the conditions more attractive and was happy to play a good deal of cricket. What has this got to do with fairness? In cricket, many decisions are made by umpires. When I was batting, there were occasions when the umpire decided that I had been dismissed (I was out) when I knew that the decision was wrong. On occasions, some decisions are difficult for an umpire to get right. For example, if the ball finds the edge of the bat and is caught, then the batter is given out. Sometimes, it is difficult to know if the ball has been snicked or if it has hit the pad or some part of the body on the way through. What does one do if given out unfairly? There is nothing one can do. The umpire's decision is final although nowadays, it can be overturned by watching a replay that has been photographed. When I was playing, this technology was not in use. When you are given out unfairly by the umpire, all you can do is suck it up and not dwell on it. There were also occasions when an umpire gave me not out when I knew I should have been given out. Life is a bit like that. You are dealt a bad hand in a game of poker, you just accept it gracefully and get on with the game.

Gullibility Is Widespread

While on the topic of cricket, I will digress to recount a humorous incident. During my time in England, I worked in a large research laboratory. It was decided to form a cricket team to play matches with similar institutes.

At the first meeting to organize the team, someone turned up with his own bat and gloves, looking every bit an accomplished cricketer. He (K.E.) was duly elected captain. In the first few matches we played, K.E. was "unlucky" in being dismissed early without making a score. On one occasion, after he had failed again, one of the team remarked, "The opposing team was lucky, K.E. didn't get going today". Some members of the team vigorously nodded their agreement. In a later match, I was at the other end when K.E. came in to bat. The bowler had twisted his ankle and had to be replaced. The youngster who was called on as a replacement was not really a bowler. He bowled the first ball to K.E. and, as I recall, I think it bounced two or three times before it reached the batsman. It reminded me of a film "The dam busters" when Allied planes attacked the great dams in the Ruhr Valley in World War II by projecting bombs that bounced along the water before exploding at the dam wall. K.E. made a cross-batted swish, missing the ball which went through to hit the stumps. It was one of those moments frozen in time. "Right", I thought "this confirms what I suspected". I recount this anecdote to illustrate a pattern that I have often observed. A person is able to achieve a reputation that has no basis but is reinforced by those who go along with what others say.

This may seem an irrelevant digression from the main themes of the book. However, one of the themes of the book is that a large number of the population behave as "useful idiots". They don't have the capacity to critically evaluate an issue and are easily deceived into believing what they are told without having been given adequate evidence. In legal terms, this is referred to as hearsay. A current example of this is the belief shared by many that human activities are having a large effect on climate and that strong actions are needed to arrest the trend. They adopt the belief without any real evidential basis. They may be influenced by hearing or reading what they perceive as authoritative opinions. They may experience events that they have not previously experienced such as floods, heatwaves, or bush fires and are induced to thinking that the climate is changing when, in fact, it is simply weather aberrations and not permanent changes. This behaviour illustrates how a belief, once adopted by a majority of the public, is almost impossible to change, even though rational arguments showing it to be false are expounded.

We Have to Accept That There Will Be Unfairness

If we focus on unfairness, we see examples of it everywhere. Some unfortunate individuals are struck down by illness, some are victims of traffic accidents which may leave them disabled, some lose their livelihood or their wealth through events outside their control. Apart from these natural misfortunes, however, a perception of unfairness seems to have crept into society. Instead of accepting the hand you were dealt, there is a mindset among some that one is a victim of an unfair society. Of course, we should never accept injustice and should oppose it wherever possible. But to obsess with the "slings and arrows of outrageous fortune" over matters that are relatively trivial or that one cannot change is not worthwhile. One has to accept that there will be bumps in the road, some bigger than others and that the bumps encountered by some people will be harder to overcome. O.K. bumps will always be there but it is best to put them out of mind and move on. Nothing is gained by dwelling on misfortune. It is much more beneficial to discard negativity and to concentrate one's mind on trying to foster a positive and optimistic attitude. Some advice on how to achieve this will be given in Chapter 6 when we consider the work of Martin Seligman.

Reference

Thompson, B. 2018. Gina Rinehart pays more tax than any other Australian. Financial Review. November 1. www.afr.com/companies/gina-rinehart-says-she-pays-more-tax-than-any-other-australian-20181101-h17d56

chapter six

Education, Schooling, and the Curriculum

When people discuss issues rationally and are prepared to concede the merit of opposing viewpoints, this represents one of the highest achievements of a civilized society. Unfortunately, this idealized situation is not always attained. Thus, we have movements where the participants who discuss an issue are certain that they have the correct viewpoint. It follows logically, in their minds at least, that anyone who doesn't agree must be wrong. What can result is discord and possibly violence. Adoption of the true scientific method can help to prevent this undesirable outcome. It requires putting aside emotional reactions and adopting reason. It doesn't mean giving up strong feelings but to learn to be detached from them when arriving at decisions or debating issues. It is an approach that requires discipline and constant practice until it becomes innate. How good would it be if this were inculcated more in our education system instead of the brainwashing that is so prevalent in our schools these days!

Another attribute that young people would benefit from acquiring is resilience. Development of this quality requires a certain amount of struggle, facing difficulties and striving to overcome them. Unfortunately, in today's world, notably in prosperous Western countries, comfortable lifestyles have become entrenched and this hinders development of resilience. Those who become resilient are more likely to develop another important human quality, that of compassion. Resilience and compassion are just two but an important two of the human qualities that are essential in formation of a successful society.

Education, in addition to science, is an area in which I have been involved and feel obliged to make some comments. Standards of education appear to have been falling in Australia (as well as some other Western countries) over the past 20 years or so. This is based on results, among others, from the OECD's Programme for International Student Assessment (PISA) which has been discussed in more detail in Chapter 3. There are three areas in which students are assessed. In this programme, these are language, science, and mathematics.

In Australia, the national curriculum has been developed by ACARA, the Australian Curriculum Assessment and Reporting Authority. ACARA has a website which provides valuable information for teachers, parents,

DOI: 10.1201/9781003254065-6

students, and the community on guidelines for what students should learn. The curriculum has been developed through an extensive and rigorous process. It includes wide consultation with experts in the area. It also uses comparisons with countries that are considered to be high performers as measured by education rankings. These include Canada (especially the province of British Columbia), Finland, Singapore, and New Zealand.

The current national curriculum for Australian schools which was introduced in 2014 includes eight learning areas. These are English, Mathematics, Science, Health and Physical Education, Humanities and Social Science, the Arts, Technologies, and Languages. Three cross-curriculum priorities have also been introduced. The three priorities that are presently included are as follows: (1) Aboriginal and Torres Strait Island histories and cultures, (2) Sustainability, and (3) Asia and Australia's engagement with Asia. The ACARA website explains the cross-curriculum priorities as follows:

> The cross-curriculum priorities are embedded in all learning areas as appropriate. When planning teaching and learning programmes for the Australian Curriculum, teachers will notice that the three cross-curriculum priorities have a strong but varying presence depending on their relevance to the learning areas.

The rationale for the choice of the three cross-curriculum priorities does not seem obvious. It does appear that they have been chosen, based on ideological considerations. This becomes apparent if we compare each with priorities that could have come from the opposite side of politics as is done in the table.

Current Cross-Curriculum Priorities	Alternative Cross-Curriculum Priorities
*Aboriginal and Torres Strait Islander histories and cultures	British heritage based on separation of church and state, parliamentary and judicial systems, Judeo-Christianity
*Asia and Australia's engagement with Asia	Europe and Australia's engagement with Europe
*Sustainability	Utilization of resources for maximum economic benefit

The curriculum was reviewed by two experts, Dr. Kevin Donnelly and Professor Ken Wilshire after inviting submissions from the public. The review was presented to the federal government. It raised many important issues that invited actions to improve the education system. Unfortunately, the report from the reviewers seems to have been completely ignored. The review of the curriculum itself will not be discussed in detail here except to quote one comment: "The reviewers heard substantial evidence that content was added to the curriculum to appease stakeholders, which has led to an overcrowded curriculum". I would like only to focus on the cross-curriculum priorities. The following were some of the concerns expressed about them in submissions to the reviewers:

1. Expertise requires deep knowledge of a particular discipline and does not readily transfer across disciplines.
2. The subject discipline must not be force-fitted to subject curriculum content when, in reality, there is no pre-existing connection.
3. The rationale for the three priorities is largely political.

The rationale for each of the three priorities is outlined on the ACARA website. I will make just one comment on each of the three priorities:

It is stated that "Aboriginal peoples and Torres Strait Islander peoples have worked scientifically for millennia and continue to contribute to contemporary science". Certainly, these peoples have developed useful knowledge which might be included in Humanities and Social Science. However, it is very different to what is understood as science in Western culture. Science in Western culture incorporates the great advances that have been made in physics, chemistry, botany, zoology, geology, etc. and the applied sciences that have emanated from these fundamental ones.

For the topic of Sustainability, the term "social justice" is mentioned five times in a relatively short section aimed at justifying its choice. This is patently a political term. In relation to Asia and Australia's engagement, it is stated that "It reflects Australia's engagement with Asia in social, cultural, political and economic spheres". This raises questions which are not clearly addressed on how exactly Australia is to engage politically with countries having diametrically opposite political systems such as the People's Republic of China, Lao Peoples Democratic Republic, and the Socialist Republic of Vietnam.

As emphasized in other parts of the book, a successful democracy involves a contest of ideas from both left and right sides of the political spectrum. It is not the intention here, as far as possible, to favour one side. However, if there is an imbalance of ideologies in public policies, this needs to be pointed out and should, if possible, be corrected. The political views of students ideally should be allowed to develop, free of influences from teachers who try to impose their beliefs. The cross-curriculum priorities that have been introduced into the Australian national curriculum clash with this aim and favour left-leaning issues. Education should ideally be free from imposition of any particular ideology. Otherwise, this can lead to brainwashing. The tendency to impose political viewpoints on impressionable young students has, unfortunately, become evident in Western countries. This has been pointed out by Katie Hopkins (2017), who stated that "schools are supposed to teach kids HOW to think, not WHAT to think. So why are many liberal teachers bullying and brainwashing children with their own intolerant views?" Similarly, Caroline Marcus (2016) asks, "Why are Australian kids becoming dumb and dumber? Ask the teachers". She attributes it to teachers wasting time on ideological brainwashing instead of focusing on literacy and numeracy.

The standard of thinking by members of a society at a particular time can, to a large extent, be traced back to influences from the education that they received in their formative years. It is therefore important that students should receive instruction that is free from ideological coercion. The aim of education should be to develop a society made up of free thinkers.

This means a society whose members can form opinions free from indoctrination. The former Russian leader Vladimir Lenin is quoted as saying "Give me just one generation of youth, and I'll transform the whole world". Another statement attributed to him was: "Give us the child for eight years and it will be a Bolshevik forever". These sorts of statements infer that a generation of people can be persuaded into believing in what one individual deems to be the ideal Utopian state.

Such assertions are made by people who feel certain that their virtue is beyond question and therefore they must have the right viewpoint. The scientific approach is different. True scientists form opinions but they are never considered to be beyond doubt. They are always open to being challenged in debate and are ready to see the merit in alternative arguments. The contrast between the two ways of thinking is striking. It is ultimately the criterion that distinguishes a free society from a totalitarian one. What happens if someone believes in a certain ideology and is never prepared to admit that it could be erroneous? What happens if there are flaws in the ideology? Since you can never admit to being wrong, you are then forced to stubbornly defend it for there is no alternative path. The scientific method, by contrast, is based on the principle that we always retain doubt and are disposed to admit our mistakes and to learn from them. That enables the possibility for progress.

Ideas of John Gatto

Based on these considerations, how should we approach the question of how best to educate our children? Some cogent arguments to address this question are made in the book "Dumbing us down. The hidden curriculum of compulsory schooling" by John Taylor Gatto (Gatto, 2005). Gatto was a school teacher in the U.S.A. for some 30 years so he was well qualified to comment on the subject. He begins the book by outlining what he perceives as mistakes that are being made in education of young people. This fits with the title of the present book: Learning from our mistakes. A distinction is made between education and schooling. The present compulsory school system is not considered to be the ideal one. Gatto found that the children he taught had almost no curiosity. He believed that a debate is needed to consider better ways to educate children. This is a debate that he believes should be repeated day after day and year after year until either the present system is fixed or, if not, alternative systems should be pursued.

The present system of mass schooling in which children are imprisoned for several hours each day tends to produce conformists instead of unique individuals. Throwing more money and more people at the problem, as many politicians urge, will only make it worse. In many countries, control of the education system is centralized. As a result, many entities depend on such a system. Gatto details some of them – analysts, advisors, textbook committees, teacher's colleges, state departments of education, and other school-related businesses. Gatto refers to a "parasitic growth of the government monopoly over the school concept".

The central theme in this book is identifying mistakes, recognizing them as mistakes, and adopting policies to correct them. John Gatto has identified what he perceived as mistakes in the education system. Some of these are the negative effects of central control with commands coming from "experts", a central elite of social engineers, imposition of conformity, and prioritizing preparation for economic success instead of fostering creativity. The system doesn't work because it is mechanical, antihuman, and hostile to family life. Gatto draws a distinction between "networks" (which include schools) and communities (including families) that have more meaningful relationships.

He illustrates the point by referring to what happened in the colonial New England states of the U.S.A. during the 1600s. Congregations took on the responsibility for solving their own problems rather than submitting to some authority. This involved a type of dialectic thinking. This refers to the ability to look at issues from multiple perspectives and through honest argument, to arrive at an optimum position. For example, a community may comprise individuals all with different opinions on how to proceed. Some may feel a priority to construct a children's playground, others may have a preference for developing a bowling green. Although they have different views, they share a common goal which is to achieve the best for everyone. Through argument and compromise, they are able to progress towards a harmonious community. There is a sharp contrast to the situation where there is central control. Those who are in control are experts who "know that they are right" and impose what they believe the community needs.

I have observed the superior nature of communities. As I described in Chapter 1, I was born and grew up in a small rural town. It has been over 40 years since I left the town but have been able to retain some contact. This has been mainly through having access to a local newspaper which is regularly produced. Residents have been progressive by keeping up to date with advances in technology and in other ways. The community, made up of like-minded people, has prospered through volunteers generously giving their time to participate in cultural activities that enrich the lives of everyone. I believe this is an example of many communities throughout the world that John Gatto was referring to. In contrast, I have witnessed actions by governments and bureaucracies (more akin to Gatto's networks) that have made societies worse rather than better. For example, they have brought into the country people who do not share the same values and will never assimilate. It is justified by appeals to a naïve interpretation of inclusiveness.

Ideas of John Medina

John Medina, a development molecular biologist, is another who has questioned the present system of schooling and pointed out some of what he believes to be its flaws. He has applied his scientific knowledge to the classroom situation. His work appears in a number of books (e.g. Medina, 2018) as well as talks and films available on the internet (www.brain-rules.net/dvd). He is well known for his book on brain rules (Medina, 2014), a New York Times Bestseller. Medina readily admits that there is much about how the brain works that we don't understand but that there are some things that have been found and we should make use of that knowledge. The first of the 12 rules is that exercise boosts brainpower. Movement has been a feature of human evolution and is found to improve thinking skills. He thus believes, as did John Gatto, that the sedentary life that school students are subjected to for hours during each day is not the most appropriate system.

Medina declares that there is no greater anti-brain environment than a classroom or a cubicle. He suggests that the classroom set-up could be modified to allow more physical exercise. For example, treadmills could be incorporated. More effective teaching might be achieved by alternating aerobic exercise with periods of focused learning. This idea could be tested by varying the sequence of physical activity and learning with the aim of discovering the optimum balance.

Ideas of Martin Seligman

Martin Seligman is an American psychologist, noted for his books on self-help. He is known for his theory of learned helplessness. He and colleagues have studied how people respond to trauma in life. They found that some (humans as well as other living beings) react to a difficult situation by resigning themselves to failure. When confronted by further traumatic events, they passively accept that they are failures (learned helplessness). However, others who encounter trauma, although first becoming depressed, are able to recover and overcome their failures. Seligman and his team made an extensive study of the different behavioural reactions. Some of the studies involved the military, a cohort who are subjected to the greatest trauma. For some, particularly those who had engaged in armed conflict, the result would often be post-traumatic stress (PTSD), depression, and suicide. Others, however, were able to deal with the issues and overcome their effects. As a result of their studies, the team were able, by means of tests, to distinguish between two groups, those who gave up (first group) and those who were able to put the traumatic experiences behind them (second group). Further, they developed courses in which they were able to train members of the first group to change their response so that they could move into the second group.

The different ways that humans react to extreme adversity can be represented by a normal distribution curve. Those at one end fall into helplessness after a traumatic event while, at the other end of the distribution, there are those who not only recover from the trauma but go on to become more resilient than before. This behaviour has been described by the great philosopher Friedrich Nietzsche who said "That which does not kill us makes us stronger". The key to the different behaviour of the two groups was optimism. The training involved techniques to reject feelings of pessimism and to replace them with feelings of optimism. Seligman developed PERMA, a model with five core elements for psychological well-being. These are **P**ositive emotions: feeling good, pleasure, and enjoyment; **E**ngagement: fulfilling work, interesting hobbies; **R**elationships: social connections, love, intimacy, emotional and physical interaction; **M**eaning: having a purpose, finding a meaning in life; **A**ccomplishments: ambition, realistic goals, important achievements, pride in yourself. Books authored by Seligman as well as YouTube interviews can be found on the internet, e.g. Seligman (2006).

Teaching of History

Balanced teaching of a nation's history should be a fundamental requisite of education. The achievements as well as the failures need to be presented equitably. In Australia, there seems to have been a distorted view of the nation's history. We see young people pushing agendas that denigrate the nation. The evils of colonization, the unfair treatment of the indigenous population, and the over-exploitation of resources are some of the issues that are being highlighted. These are based on the political concept of "social justice".

It is proper that mistakes be recognized. It is also proper that, in teaching history, there be a balance between the failures and the achievements. Presently, this balance is lacking. Negative policies in the past should be exposed but not to the extent of creating resentment and a mentality of vic-timhood that is happening in some sections of the community. Obsession with mistakes creates negativity. We can't change the past. It is better to admit the mistakes but to learn from them to enable us to plan a course for a fairer and more optimistic future. Educators should also acknowledge the great accomplishments that have been made. This will help to foster the national pride and patriotism that is needed for a coherent society.

Critical Thinking

The ability to think critically should be a priority in education. There are many books available that deal with critical thinking. Each one contains useful information. I would like to discuss the subject from a direction that has not received a lot of coverage. This is critical thinking based on the scientific method. As has been discussed previously, the hypothesis-deductive method is what I believe to be the truest procedure for advanc-ing knowledge using science.

Those who are skilled in critical thinking are able to discern faults in thinking when they appear. The aim in education should be to help students to become proficient in recognizing false thinking. The scientific method is a relentless pursuit of truth. It never claims to reach certainty but does have criteria for knowing how to approach truth. When we pick up a newspaper and read an article, we should reserve judgement until we have seen what is written. However, on regularly reading a particular writer's articles, it can be anticipated that it will be biased in a certain direction. In each of our minds, there is an ideological prism embedded. When facts are received by the mind, the prism deviates the information so as to accord with the ideological bent. For some, particularly those who have developed critical thinking skills, there will be little deviation– the information comes straight through without much change. For others, the facts will be massaged, some not by very much but others by a great deal. The ideological prism (admittedly an abstract concept) is a deceptive device. A person may be unaware it exists and believes that what they say or write is simply the truth.

Understanding the Scientific Method

In the public debate, we regularly hear statements such as "It has been scientifically proven". This shows a misunderstanding of the scientific method. This misunderstanding extends to national leaders. It means that policies that are claimed to be based on science can be mistaken and the mistakes can prove costly as is explained in other parts of the book. Therefore, there is an urgent need for sound instruction on the correct phi-losophy of the scientific method to be included in early education. There should be an explanation of some of its most important concepts. These include the hypothetico-deductive method as the correct one for advancing scientific knowledge, the limitations of induction, invalid conclusions based on correlations, selection bias, and the importance of control experiments.

References

Gatto, J.T. 2005. Dumbing us down: the hidden curriculum of compulsory schooling. New Society Publishers, Gabriola, B.C. Canada.

Hopkins, K. 2017. Schools are supposed to teach kids HOW to think for themselves not WHAT to think. So why are so many liberal teachers bullying and brainwashing children with their own intolerant views. Daily Mail, February 6. www.dailymail.co.uk/debate/article-4194048/KATIE-HOPKINS-liberal-brainwashing-schools.html

Marcus, C. 2016. Why are Australians becoming dumb and dumber? Ask the teachers. Daily Telegraph, Melbourne, December 20.

Medina, J. 2014. 12 Principles for surviving and thriving at work, home and school. Scribe Publications, Brunswick.

Medina, J. 2018. Attack of the teenage brain! Understanding and supporting the weird and wonderful adolescent learner. ASCD, Alexandria, VA.

Seligman, M. 2006. Learned optimism: how to change your mind and your life. Random House USA Inc., New York, NY.

chapter seven

Where Have We Been Going Wrong in Our Society?

The title of the chapter asks a question which we will attempt to address. The question may be criticized by some. They may ask "On what basis do you assume that we are going wrong?" Some may say "We are not going wrong. We are on the right path". Whether we are going in the right direction or the wrong one is a matter of opinion. The opinion that I share with many is that, in some ways, we are going wrong and some of my reasons for thinking so are alluded to in earlier parts of the book but will be expanded on here.

Before doing so, an examination of human history suggests that there has been a continuous improvement of society providing we consider a long enough timescale. This should make us optimistic and justifies a positive outlook for the future. It is the relatively short periods of retrogression that are of concern. These require actions that make corrections to enable continuation of the long-term improvement. To begin, let me outline briefly what I consider to be the right path and what I consider to be the wrong one. I will then try to justify the viewpoint. Readers will then be free to agree or disagree. The ensuing debate may then help us all to arrive at a better mutual understanding.

I consider that the right path for a society at present is one that adopts democracy while acknowledging its imperfections and striving to correct them. A democracy seems to be the political system that gives citizens the most freedom if we exclude anarchy. I believe that the best society is one that is made up of free thinkers in which people have the liberty to determine how they wish to live providing this does not interfere with the rights and aspirations of others. The wrong path is the one that creates a society made up of people who have been manipulated or indoctrinated or have been made to be subservient to follow a specific ideology. Freedoms have been curtailed and lives have been subjected to tight controls by an authoritarian governing regime. In simple terms, the argument proposed is that freedom is better than enslavement.

In Chapter 4, we looked briefly at how a successful democracy should function with inputs from the two main sides of politics. The intention in this book is to try to be impartial and not favour either side. What will be opposed, however, is if there is seen to be a move towards extreme

DOI: 10.1201/9781003254065-7

positions. We have learnt from history of the disastrous consequences resulting from movements that go too far to the left or too far to the right. In either case, this represents a movement towards totalitarianism which I believe should be resisted.

It has become evident in recent decades that there has been a change of balance in the influence that political ideology has had on Western society. Some of the examples that will be given are to Australia but they relate generally to all Western countries. This change has been manifested by imposition of left-leaning policies in different areas such as education, the media, the judiciary and the bureaucracy. It has been described as "the march through our institutions", referring to an activist movement to steer political ideology to the extreme left.

In an article published in The Australian newspaper, Judith Sloane (Sloane, 2020) has succinctly summarized some of the outcomes of the trend. She states, "The march has taken a while, led by universities veering to the left and inculcating the full agenda of woke topics – the evils of capitalism, colonialism, racism, women's rights, inequality, diversity, multiculturalism, intersectionality, climate change etc.". She points out, rather ominously, that the influences have not been restricted to humanities and law students but have been extended to medicine, science, engineering, and business. The result has been that graduates who have been groomed in political correctness have entered the workforce and are promulgating and imposing their views (and policies where they attain power) in different areas of society.

To answer the question "Where have we been going wrong?", I will draw on my knowledge of the scientific method. It first needs to be admitted that the scientific method has its limitations. It only has the capacity to give partial answers to some questions but, at least, that is a start. Issues of morals and ethics are outside the scope of science and these are issues that need to be considered when answering the question. I know a little about morals and ethics but I will leave their discussion to those who are more qualified. My guess would be that one answer could be that we have tended in recent times to ignore the great teachers of morals such as Jesus of Nazareth and many others. They may say that we have dismissed the warnings from these great teachers about heeding false prophets. This leaves us with only science to try to address the question, realizing that the answer can only be incomplete.

The basis of science is the hypothetico-deductive method of enquiry. This method has been explained in Chapter 2. To simplify it, let us think of the method in simple layperson's terms. It can be encapsulated by the saying "We learn from our mistakes". It goes something like this. We are confronted by a problem or a question. We try to resolve the problem or answer the question by tentatively suggesting a possible solution (a hypothesis). The suggested solution must be one that, in principle, can be refuted. It is then subjected to a test that must enable a possible refutation. Does it stand up to the examination or does it fail the test? If it fails, we need to ensure two things. First, we need to be sure that there are no doubts that the test was properly applied so that we have confidence in the result. If we are sure that it was, then we need to accept that the proposed solution is wrong and begin to think of an alternative hypothesis. Thinking about a new hypothesis can be stimulated by learning from the failed one. It is thus a trial and error procedure in which we learn from our mistakes.

The opposite approach has obviously to be that we don't learn from our mistakes. This may be because we believe that we haven't made mistakes. What we believe is that we know the truth. All our subsequent decisions are therefore based on this belief. There we have it. We then need to ask "Is it better to steadfastly believe what we perceive to be the truth or, as in the scientific method, to always retain doubts and never assume that we have the truth?" Many would side with the former. Isn't it better to feel sure and to have complete confidence in what you believe than to always hold doubts? But what if the belief you hold is false? Just because you have a strong dogmatic belief doesn't mean that it is necessarily true. What is the criterion for believing that you hold the truth? Rationally, there may not be one. On the other hand, true scientists, although never accepting that they know the absolute truth, do have a criterion for establishing if they are moving towards the truth.

If a hypothesis is tested and shown to be wrong, this eliminates one possible explanation. The ruling out of one explanation can stimulate further thought for forming an alternative hypothesis and for novel experiments to test it. If, after one or more failed hypotheses, severe tests of a new hypothesis are unable to refute it, the new hypothesis is said to be corroborated. Although it can never be assumed that the absolute truth has been reached, more confidence is justified that we are moving towards it. The trial and error procedure of the scientific method, therefore, serves as a criterion for approaching the truth.

Let us return to the question "Where are we going wrong?" or, put another way, "What mistakes are we making?" If we are making mistakes and have the capacity to recognize that we are, then we are in a favourable position. This is because we can apply the scientific method to learn from the mistakes and be in a position to correct them. This is the great advantage over the alternative position in which we are unable to recognize any mistakes and so continue blindly on the assumption that we are on the right path.

I believe that there are many areas where we are making mistakes. Some of these areas, such as education policies, have been alluded to in earlier chapters. The mistakes made in these more general areas will be discussed later in the chapter. First, I would like to focus on two specific areas where I believe that serious mistakes have been made in adopting beliefs and in adopting policies based on those beliefs. One is the belief that green-house gases from industrial processes are causing dangerous warming of the planet. A corollary of this is that fossil fuels should be eliminated and replaced by non-emitting renewable sources of energy. The other mistaken belief is that the presence of an infectious coronavirus warrants a complete community lockdown. The actions imposed by governments to address these two issues have caused and are causing serious economic damage and contributing to lowering of living standards, at least for some. Before dealing separately with each issue, there are similarities to be noted in how each has evolved. Neither has been based on solid science and both have encouraged governments to exploit fear to make a gullible public accept draconian measures to counter the supposed threats.

Anthropogenic Global Warming

Since the beginning of the industrial revolution, processes such as burn-ing of fossil fuels have produced what are called greenhouse gases. The temperature of the earth is regulated by its atmosphere. The gases emitted by industrial processes enter the atmosphere and could in theory add a contribution that causes a global increase in temperature. However, it has never been shown whether the contribution is significant or whether it is negligible. We see changes in weather but these are temporal. No one doubts that the climate is changing over long periods but it needs observa-tions over decades to confirm that the changes are permanent and not just natural fluctuations in weather that we are seeing. The theory of anthro-pogenic global warming (AGW) is based on an assumption that human activities are producing emissions that are making significant permanent changes which could lead to dangerous and possibly catastrophic warm-ing of the planet. One of the gases that is emitted is carbon dioxide and the amount that is emitted has been adopted as a quantitative measure of industrial emissions.

An Intergovernmental Panel on Climate Change (IPCC) was formed by the United Nations (UN) in 1988. It is made up of scientists and bureau-crats. It does not carry out research itself but prepares and distributes reports. These are based on peer-reviewed literature and the aim of the IPCC is to assess the impact and possible responses to climate change. Most research on climate change has been based on theoretical model-ling rather than empirical testing. Modelling involves the input of con-tributions from all variables believed to influence a given phenomenon. This then gives an output that purports to quantitatively predict how the phenomenon will be affected by the variables. The work of the IPCC has already been described in Chapter 4. Its reports have had a huge influence in convincing the public that anthropogenic climate change is a serious challenge and needs urgent responses to mitigate its effects.

Let us look more closely at the assertion made above that AGW is not based on solid science and that consequently there is no justification for basing policies on this assumption. Of course, in accord with the true scientific method, this opinion is not held dogmatically and we must always remain open to contrary arguments and to new results that show AGW could be significant. There have been many critiques of the theory and how it has been developed but most of the criticisms have been pointed out by Kelly (2019) and these will be summarized here.

1. The official mission of the IPCC which is stated on its website is to provide governments with scientific information that they can use to develop climate policies. Note that the goal is not about understanding or evaluating climate change but about climate policy. Kelly asks "If the IPCC were to advise the governments that human activities were not causing climate change and that therefore there was no need to do anything, what would be the point in funding it?" This shows that there is a conflict of interest in its work.

2. The reports of the IPCC are not based on its own research. They are based on a consensus of a large number of scientists who work in the area. Consensus is not a criterion for the validity of a theory. Frequently, major scientific advances are the result of inputs from scientists who do not follow consensus.

3. Most of the research reported by the IPCC is based on predictions from models and not on experimental testing. These models have consistently overestimated future temperatures. Based on the philosophy of the scientific method, this should be accepted as a refutation of the theory (Popper, 2002).

4. Different models show variations and variance among models is used by the IPCC to quantify probability. For example, on this reasoning, if nine out of ten models predict human-caused warming, it is claimed the probability that there is warming is 90%. This is not a scientifically valid conclusion.

5. Climate models that are the basis of the IPCC reports say that CO_2 levels in the atmosphere are increasing so the temperature should keep rising. There has, however, been little warming since 1990. Climate researchers are therefore searching for an explanation. One that has been suggested is that aerosols in the atmosphere (e.g. smog in China) are causing a cooling effect. It is always possible to modify a theory by introducing auxiliary hypotheses but eventually, if this procedure is followed, a theory can become unfalsifiable and therefore unscientific.

6. Both temperature and CO_2 levels vary from place to place and from one time to another, making estimation of global average temperatures a speculative exercise.
7. Other variables that can affect climate are not assigned proper coverage. For example, cosmoclimatology suggests that climate has little to do with CO_2 and factors such as solar activity, cosmic rays, and clouds are more important.

The seven points taken from the article by Kelly may appear to be an overkill but each is a valid criticism of the AGW theory. If we apply the hypothetico-deductive scientific criterion (Popper, 2002), we should need only one refutation of a test to reject a hypothesis. Einstein's theory of general relativity was tested by measuring the deflection of light from distant stars when it passed near the sun during a total eclipse. The experiment confirmed that light was deflected and the magnitude of the deflection agreed with the prediction. If it had not, the theory would have been refuted. Point no. 3 was that the magnitude of warming did not agree with what was predicted from the models. Of course, it may be argued that the inputs into the models may have been flawed. In any case, it is a refutation so that until there is agreement, there is no justification for accepting the theory of AGW. If adjustments to the models are made so that they predict the warming more accurately, we then have to consider point no. 5.

True science is carried out by scientists who publish their results and theories in scientific journals or present them at conferences. The work of each individual can then be subjected to criticism by fellow scientists. In that way, mistakes can be identified in debate and knowledge can be advanced. When scientists pass their work on to bureaucrats who then try to arrive at a consensus from a large number of contributions, this is not how science is supposed to work. Such a procedure cannot be justified as a method for approaching the truth. Although it is widely presented as "the science" and swallowed by many of those who determine policies as well as a large proportion of the general public, it is not truly science.

It therefore seems remarkable that AGW has been widely accepted as authentic when it is not supported by true science. The media has been particularly influential in convincing an otherwise uninformed public. We regularly see articles in the press with alarming headings such as "New evidence shows that climate change is worse than was thought". Many young people have been indoctrinated to believe that planet Earth is heading for catastrophic warming. A rumour has spread, fomented by the media that the world was going to end in 2030 unless drastic action was taken on climate change. It seems to have been based on a misrepresentation of the 2018 IPCC report. This report simply said that to have a good chance of limiting warming to 1.5°C from pre-industrial times, CO_2 emissions needed to decline 45% by 2030. This declaration was based on unsubstantiated models. The conclusion in itself is fanciful; as if humans were capable of dialling down the temperature by a predetermined amount. In Australia, children have wagged school to participate in protests. The protests are organized to admonish politicians for not taking sufficient action to avoid the assumed future crisis. The UN has invited a schoolgirl to address the Assembly on the matter. In contrast, scientists who are knowledgeable about the topic are attacked for not sharing the politically correct viewpoint. For example, Ivar Giaever, a Norwegian-American physicist and a Nobel Laureate, is branded as a climate sceptic and not a climate scientist. The term climate sceptic is intended to have a negative connotation even though all scientists should be sceptics. Another erroneous implication is that only those who are considered to be "climate scientists" can give opinions that have credibility.

It might be opportune at this point to make a relevant comment. The fundamental sciences include physics, chemistry, biology, geology, and mathematics. Then there are the applied sciences such as food science, agricultural science, and climate science. An applied scientist will have knowledge of a number of sciences. For example, a climate scientist needs to have some knowledge of physics, chemistry, meteorology, and oceanography. However, an applied scientist will usually not have the depth of knowledge of one science that a fundamental scientist has. It is mainly depth of knowledge that is needed to make genuine scientific advances.

We have critically examined the theory of anthropogenic climate change and pointed out some of its flaws. In view of this, what I believe needs to be asked is how the theory has gained such wide support and formed the basis for government policies that are having huge detrimental effects. To answer this question, we need to focus on how it originated. A paper currently being reviewed but publically available, written by the Argonauts, a team of self-funded scientists, has addressed the question. This group which includes eminent climatologists and control theorists points out that the climate "emergency" originates from a grave error of physics. As it is a 70-page paper, only a brief summary of it will be given here.

Before doing so, we need to understand an important topic that impinges on climate science. It is the concept of feedback. A feedback process is a secondary effect that results from an initial effect that, in the case of climate, is referred to as a climate forcing. Climate feedback processes can either amplify or diminish the effect of climate forcing. An example of a positive feedback is when emitted greenhouse gases increase, the atmosphere warms, leading to more evaporation of water vapour (a greenhouse gas) from oceans. Warm air holds more water vapour which amplifies the initial warming. This, in turn, causes more evaporation, leading to the cycle being continually repeated, a process referred to as a positive feedback loop. A negative climate feedback can occur when the increase in temperature causes an increase in cloud cover. This reduces incoming solar radiation and thus reduces warming.

The Argonauts have pointed out two errors in the original calculations by climate scientists of the warming that would occur from industrial emissions. These errors have resulted in overestimates by a factor of three in the predicted temperature increase, considering the period from 1850 (start of the industrial revolution) to the present. The first was the failure to take into account that, without greenhouse gases in the air, there would be no clouds to reflect solar radiation back to space. The second mistake, a larger one, was the assumption that warming in the pre-industrial period was triggered by the noncondensing greenhouse gases (gases other than water vapour) when in fact it was due to the sun. The climate scientists who were responsible for initiating the alarm about the effects of human influences had added this solar feedback response and miscounted it as part of the pre-industrial greenhouse gas feedback response. The result, due to the error, was to predict an AGW three times too large.

When corrections to climate models are made to eliminate the errors, the predicted global warming agrees more closely with what is being observed. The authors from the Argonauts state,

> There will be too little global warming to harm us. Small, slow warming will be a good thing overall. There is no climate emergency. There never was. The trillions wasted on destroying jobs and industries can now be spent on the world's real environment problems. Global warming is not one of them.

The belief that dangerous warming results from emissions by fossil fuels has been strongly promoted by the UN, particularly through its IPCC. It has culminated in the formation of the Framework Convention on Climate Change. This is a legally binding international treaty, signed by 196 parties in Paris on December 12, 2015 and entered into force on November 4, 2016. Its goal is to limit global warming to well below 2°C, preferably 1.5°C compared to pre-industrial levels.

Craig Kelly, an Australian parliamentarian, has described how the initial euphoria about the agreement has died down as many of the signatories have recognized the problems it has raised (Kelly, 2019). He points out that, under the agreement, "Western nations have agreed to punish their economies, limit their growth and spend tens of billions in unnecessary costs – all in the belief that this will somehow reduce the incidents of bad weather". In contrast, the Chinese Communist Party will not be held to the same obligation. This will allow it to expand its strength, economically, politically, and militarily to the detriment of Western democracies.

What is behind these initiatives by the UN? The organization, although supposedly defending the interests of all nations, appears to be stacking its leadership with individuals who are supporters of socialist ideology. The policies being introduced are made at gatherings held in attractive venues where authoritative leaders together with their followers meet to decide policies that are increasingly being shown to be based on falsehoods.

Acceptance of the belief that emissions from industrial processes are causing dangerous climate change has led to policies to phase out fossil fuels such as coal and gas that have been the main traditional energy sources. To substitute for the resulting energy deficit, alternative energy sources such as wind, solar power, and storage batteries are being developed. These sources not only are non-emitters but have the advantage of being renewable, unlike coal and petroleum which have finite lifetimes. In order to assist their development, subsidies are being apportioned to these renewable sources. The net effect of these policies has been to increase electricity costs to everyone. Apart from the higher cost, there have been other unintended consequences:

1. Wind only produces power when it blows and the sun only supplies power when it shines. These sources of energy are therefore intermittent. They cannot supply enough power when it is needed and too much when it is not needed. They therefore need to be supplemented by reliable sources such as coal-fired power to maintain a constant baseload. Some coal power stations in Australia have been destroyed based on the assumption that they would not be needed in the future.

2. Although touted as being environment-friendly, the manufacturers of wind turbines and solar panels utilize materials that can be harmful. Slave labour is being used to mine the lithium and cobalt needed for solar cells. Both wind turbines and solar panels have limited lifetimes which leads to the problem of hazardous waste disposal. For example, solar panels contain lead and carcinogenic cadmium. When solar panels are disposed, these pollutants can be washed out over months by rainwater.

3. Wind turbines and solar panels require large areas of land if they are to produce the energy that has been supplied by coal. The sound emitted by rotating blades of wind turbines has been found to have harmful effects on bird life as well as human health and court cases have been initiated in response to complaints. Wind turbines spoil the landscape and cause destruction of wildlife, especially large birds.

The policy of developing alternative energy sources to replace traditional ones therefore raises questions. Has it been a mistake? We must be constantly questioning whether we are making the right decisions.

Complacency is not an option. The rationale for the replacement is based on the presumption that emissions from fossil fuels are causing dangerous global warming. But this has not been scientifically corroborated as discussed earlier in the chapter. Thus, have mistaken policies been adopted to address an assumed problem that is itself mistaken?

The Lockdowns in Response to the Coronavirus

The SARS-CoV-2 virus originated in China and spread rapidly throughout the world from early 2020. At the end of October 2020, the number of cases was reported to be 48 million of which 34.7 million had recovered. There were 1.23 million reported deaths although these were people who died with the disease and not necessarily from it. The test used to detect the presence of the virus is a polymerase chain reaction (PCR) test. This test uses cycles of amplification of genetic material. It is thus a non-specific test in which contamination by other genetic material can give rise to false positives. The constant reporting by the media of the number of cases measured by the PCR test has alarmed the public. It has induced governments to take steps to reduce transmittance of the virus by imposing lockdowns. Most cases have led to relatively mild symptoms. The deaths have been limited to people of advanced age (over 70) and those with other health conditions. Younger people have not been greatly affected.

In spite of this, a total lockdown has been mandated in many countries. This has involved restricting movement of people from their homes, closing of businesses such as restaurants and, in some cases, mandatory wearing of face masks in public. In Australia, the government has compensated those who have lost their jobs by schemes in which employees are paid via their employers to nominally retain their employment. The net result has been a severe hit to the economy.

The issue to be discussed here is that the decisions for lockdowns in response to the virus have not been based on solid science. These decisions have been subjected to criticisms from many sources. Perhaps the most clearly enunciated criticism has come from an open letter sent by a group of Belgian medical doctors and health professionals to Belgian authorities and media (DOCS4 Open Debate, 2020). The letter submits that the collateral damage from the lockdown will have a greater negative impact in the short and long term on the whole population than the benefits achieved from measures put in place to safeguard the population. These medical experts assert that the policies that have been introduced (e.g. lockdowns) have not been sufficiently scientifically based. They are also critical of the lack of opportunity for an open debate in the media in which different viewpoints could be debated. Instead of this, policies have been mandated without sufficient thought and consultation. At the beginning of the pandemic, the need to take definitive precautions was understandable. The

World Health Organization (WHO) predicted that the pandemic would claim 3–4% victims; i.e. millions of deaths, putting unprecedented pressure on intensive care units (ICUs) of hospitals. Such prognostications (which did not eventuate) were part of the rationale for lockdown of the entire society in some countries, including quarantining of healthy people.

The Belgian specialists asserted that the course of the virus infection followed the course of the wave of infection encountered in a normal influenza season. They criticized the application of the PCR tests and the justification for imposing measures based solely on these tests. It was pointed out that when comparisons are made with countries that did not mandate lockdowns (e.g. Sweden), similar curves for the number of cases were obtained. Lockdowns did not lead to a lower mortality rate. In scientific research, one of the strategies is to perform control experiments. This involves carrying out an experiment in which a variable is introduced and comparing the result with the same experiment in which the variable is omitted. The work on the coronavirus has the advantage that control experiments are available. This has shown that lockdowns did not lead to a lower mortality rate.

The Belgian specialists explained clearly how an immune system works. A strong immune system relies on normal exposure to microorganisms. Overly hygienic measures have a detrimental effect on immunity. What can be taken from this is that people with weak or faulty immune systems should receive special care. It includes elderly people and those with pre-existing health problems. However, providing care is taken with this cohort, it is counterproductive to subject younger healthy people to lockdowns. These people will usually have developed cross-immunity; i.e. they have been exposed to variants of the same virus. Social isolation and economic stress can lead to increase in depression, anxiety, suicide, intra-family violence, and child abuse.

Natural immunity is a better prevention approach when combined with good nutrition, exercise in fresh air without a mask, stress reduction, and nourishment of emotional and social contacts. This has been borne out by results from second waves of infections in which there were a large number of infections but a much lower mortality rate. A therapy that has been found to be safe is the use of hydroxychloroquine (HCQ) together with zinc and azithromycin (AZT) if administered in the early stages of infection. In some countries, there has been a ban on this treatment but this appears to be unjustified. Another drug that has been claimed to successfully treat the coronavirus disease (COVID-19) is ivermectin (Capuzzo, 2021). Currently, there is a controversy in relation to its use and, like HCQ, it has not been approved in some countries. A thorough unbiased investigation is needed to discover its efficacy and the reasons for resisting its availability.

Another statement that incorporates similar points to those made by the Belgian authorities is the Great Barrington Declaration (www.gbdeclaration.org). This declaration was drafted on October 4, 2020 at the American Institute for Economic Research in Great Barrington, Massachusetts, U.S.A. It is supported by signatories of infectious disease epidemiologists and public health scientists as well as members of the general public. It calls for individuals at significantly lower risk from the virus to be allowed to resume their normal lives.

In contrast to this advice, contrary arguments have been presented in the John Snow Memorandum which was published in The Lancet on October 16, 2020. This memorandum stresses how spread of the virus was restricted by lockdowns imposed early in the detection of the disease. It acknowledges the widespread disruption that has been caused but submits that policies aimed at development of herd immunity are fallacious and not based on scientific evidence. The statement has been supported by thousands of scientists, researchers, and healthcare professionals.

We therefore have opposing views on the efficacy of lockdowns, each supported by large numbers of knowledgeable specialists. It is good that there is support for opposite opinions about lockdowns. This is how science should work. As time goes on and evidence accumulates on the nature of the virus and the efficacy of different approaches to counter it, a clearer picture of how best to deal with the disease should become clearer.

An article by Bagus et al. (2021) draws attention to the phenomenon of mass hysteria. This is a condition in which a perceived threat, in this case the COVID-19 pandemic, can cause collective anxiety and panic in a society. The authors examine how it occurs, which factors contribute to its development and the policy errors that can follow. Only a few of their findings will be mentioned as this is an ongoing area of research. The development of information technology, particularly the use of social media and the decline of religion can make a society more vulnerable to mass hysteria. The greater and more centralized a government is, the greater the damage it can inflict, especially when combined with a sensationalist media. It can result in loss of human rights such as curfews, lockdowns, and closure of businesses that result from government responses. Mass hysteria can lead to huge health costs due to psychological stress as well as indirect costs resulting from alcoholism, suicides, damage from deferred treatments and delayed prognostication of illnesses. Policy failures due to the effects of mass hysteria inevitably lead to economic decline and poverty.

Resilience

The topic of resilience was introduced in Chapter 1. It is an attribute that needs to be nurtured if we want to continue to maintain a society in which its members have the capacity to withstand hard knocks and to bounce back with renewed resolve. There are current doubts whether younger Australians have the resilience of the earlier generations who pioneered the country with struggle and sacrifice. Could the younger generations rise to emulate their ancestors if they were faced with problems of similar magnitude such as economic depressions and wars? Of course, the way to answer this is to put them in similar difficult situations and see how they perform. There are signs that they may not perform as well and this is due in part to their upbringing.

Claire Fox is one who summarizes the views of many in her article entitled "Generation Snowflake; how we train our kids to be censorious cry-babies" (Fox, 2016). She states in her article,

> Speaking at school and university events in recent years, I've noticed an increasingly aggrieved response from my young audience to any argument I put forward that they don't like. They are genuinely distressed by ideas that run contrary to their worldview. Even making a general case for free speech can lead to a gasp of disbelief. Why do they take everything so personally? Because we have socialised them that way.

Her argument is that we have denied the resilient-building freedoms that were given to earlier generations.

The term "Snowflake Generation" was one of Collins English dictionary words of the year in 2016. Collins defines the term as "the young adults of the 2010s, viewed as being less resilient and more prone to taking offense than previous generations". It has been the custom to divide the population into generations based on the periods when they were born. Thus the Boomers were born roughly in the years 1946–1965, Generation X (1966–1976), Generation Y, sometimes called millennials (1977–1994), and Generation Z (1995–2010). Each generation has been associated with certain general characteristics. This can often be unfair as each person is an individual and does not necessarily fit neatly into a box. The Snowflake Generation is considered by some to be part of the millennial generation. It is a derogatory term that is used to refer to people who are perceived to be super sensitive and intolerant of disagreement.

We are currently seeing sportsmen stopping play in a match to complain against members of the crowd who taunt them. This is a relatively new development. When I competed in sport years ago, I thrived on having abuse hurled at me. It was understood that spectators would be strongly partisan. What a weak pathetic lot we have become!

Education in the School System

Let us return to the topic of education in our school system. I have referred previously to the march towards the extreme left in our institutions. This has happened and is happening in the school systems of Western countries. There is a substantial literature on the influence of left-leaning ideology on school education. I shall only select a few examples that will illustrate the general trend. Marcus Tybalt reports his experiences in the Australian public school system before he switched to a Catholic school (Tybalt, 2017). What he reports is typical of what is happening in school systems throughout the Western world. He describes how he was exposed to programmes ranging from sex education to the stolen generation (stories about Aboriginal children being removed from their homes where they were being subjected to abuse). This type of education is most intense in the state of Victoria, Australia's most socialist state where children were being taught radical gender theory under the Safe School program. The more recent Respectful Relationships program teaches children about classic neo-Marxist concepts such as male privilege, extortion of money, and gender fluidity.

The left-wing indoctrination of students in Britain has been described by Calvin Robinson, a British teacher and commentator (Robinson, 2016). An interesting observation in his blog was that, in the BREXIT vote, which gave a choice between remain or leave the European Community which resulted in a win for LEAVE, two-thirds of young people voted REMAIN. It was inferred by many that this showed that the older generations were out of touch. Robinson, a young person himself, suggested that the case may actually have been the other way around. His point was that young people were being indoctrinated to a left-wing mentality from a very young age.

The country where the trend may be most obvious is in the world's strongest democracy, the U.S.A. It has been summarized by its former president, Donald Trump, in a speech to the White House Conference on American History at the National Archive Museum in Washington, DC (Trump, 2020). He refers to a radical movement that is attempting to destroy America's inheritance. The movement is represented by violent mobs on the street as well as the cancel culture in the boardroom. It is aimed at silencing dissent and presenting a distorted and false picture of American history. He attributes these activities to decades of left-wing indoctrination in schools.

Tertiary Education

Some of the changes made to financial assistance to Australian tertiary level students have been mentioned in Chapter 1. Prior to 1974, entry to universities was encouraged by awards of Commonwealth Scholarships based on merit and a means test. Changes in 1974 saw financial assistance made available to all students enrolled in university courses, again subject to a means test. The new system offered the possibility for everyone to enter university providing they qualified. This enhanced opportunities and, in theory, ensured a larger future cohort of educated professionals.

Have the changes all led to better outcomes? What have been the negatives? By making tertiary education available to all rather than selecting the best, meritocracy is eroded. Prior to 1974, it was probable that students entering university were more highly motivated to succeed. As it happened, the system where all students were assisted could not be sustained and fees had to be introduced. Students who were unable to meet the costs were allowed to postpone payment by taking out loans under the Higher Education Contribution Scheme (HECS). Those who graduated were burdened with repayments once their earnings reached a certain threshold level. For those who failed to complete their degrees and there were many, the government (i.e. the taxpayers) was stuck with unpaid debts that have ballooned over the years.

New universities have been created to cater for the increased number of students (see Chapter 1). Have academic standards been maintained as a result of the changes? To meet the costs, programmes have been introduced to increase the number of foreign students, who pay higher fees. This has produced a more fragile system financially. Reliance on this source of funding can leave universities in financial danger. It has happened as a result of the coronavirus pandemic which has prevented foreign students from entering or returning to the country.

Apart from the financial difficulties, other problems have arisen in tertiary education. Originating mainly in the humanities departments, there has been a progressive increase in adoption of far left thinking. There has been a push to influence students towards leftist ideology. A similar trend to what is occurring in the U.S.A. is happening in Australia. There is rejection of past achievements and greater support for socialist ideas. An illustration of this has been the reluctance of some universities to accept financial offers to develop a course in Western Civilization.

The Ramsay Centre (a philanthropic Australian organization) has been offering universities money to teach degrees in Western Civilization but has met with resistance. Concerted protests by staff and students have resulted in rejection of the proposal by the Australian National University and the University of Sydney. One of the reasons for not accepting finance is the perception that the courses may have an underlying theme of the supremacy of white civilization. Many of those who oppose it see it from a purely political viewpoint. They see its aim is to create right-wing ideologues. This is ironic considering that what has been happening in Australian universities for a long time has been the creation of left-wing ideologues. While there has been vigorous resistance to introduction of teaching Western culture, Confucius Institutes have flourished in Australian universities. Despite the initial opposition, three Australian universities have now signed up and will be offering courses on Western Civilization.

Corporations

The most coveted prize for those who support extreme left-wing policies is the ability to infiltrate capitalistic enterprises and to erode them with their long-term aim of substituting their own agendas. The objective is having some success with its influence on some large corporations

Gillette, a subsidiary of Procter & Gamble, has caused controversy by an advertising campaign based on the slogan "The best men can be". The aim is to correct negative behaviour of men such as bullying and sexual misconduct which come under the heading of toxic masculinity. There has been a backlash against this strategy which has resulted in revenue losses to the company (Baggs, 2019). The irony has not been lost on some of the general public that a company that rails against toxic masculinity is trying to sell razors to men.

Qantas, Australia's national airline, has confirmed its commitment to social issues, including LGBTI, indigenous and women's rights (Qantas, 2019). The company CEO led the airline's highly successful promotion of "same sex marriage" during the 2017 Australian plebiscite on the issue.

ANZ, one of Australia's four big banks, has issued a statement that the bank supports the Paris Climate Change Agreement's goal of transitioning to net-zero industrial emissions by 2050 (Clayton, 2020). The bank has announced that it wants to support customers and projects that contribute to reducing emissions and are resilient to climate change. It states that communities and customers could be impacted by climate change, directly through increasing severity of weather events (physical risks) or through legislative, regulatory or policy responses (transition risks).

Australia's BHP is one of the world's largest resources companies. Its activities include projects focused on fossil fuels (coal and petroleum) as well as minerals (copper, iron ore, nickel, and zinc). On its webcast (BHP, 2020), it states that it supports the aim of the Paris Climate Agreement to limit global warming to well below 2°C above pre-industrial levels and pursue efforts to limit warming to 1.5°C. It has already put in place a policy to establish its long-term goal of achieving net-zero operational emissions by 2050.

The Media

Many people are influenced in their opinions by the information that they obtain from what they hear or read in the media. In a country that is governed by an authoritarian regime, the news that is fed to the public is controlled so as not to be inconsistent with the ideology of the regime. Countries that are free are not subjected to this control. Unfortunately, this does not mean that the news presented is free from bias. The topic of media bias is succinctly analysed in an editorial from Investor's Business Daily (Investor's Business Daily, 2018). Here, the results of several surveys in the U.S.A. are summarized. It is concluded that journalism has become the most left-wing of all professions. It hasn't always been the case. The article reports a study of "The American Journalist in the Digital Age" which showed in a long-term study that, in 1971, republicans (right wing) made up 25.7% of all journalists, Democrats (left wing) were 35.5% and independents 32.5%. In 2014, journalists identifying as Republican had shrunk to 7.1%. In the 1970s, there was close to parity between the two parties but in the most recent study, Democrats outnumbered Republicans by four to one.

Despite the imbalance in political, views, the media in Western countries are free which is better than the situation in totalitarian regimes. A problem does occur, however, when the public broadcaster (i.e. funded by the taxpayer) is seen to be biased. As pointed out by Gerard Henderson (Henderson, 2020), the Australian public broadcaster (ABC) is "a conservative free zone without one conservative presenter, producer or editor for any of its prominent television, radio or online outlets". This gives one side of politics an unfair advantage as many will prefer to choose a "free" service to get their news and to have their opinions influenced.

In recent times, it has become apparent that there has been an increasing sense of ingratitude in the Australian (and other Western) populations. The sacrifices of earlier generations that have given us the improved living conditions that we all enjoy have not been fully appreciated. There has been criticism of colonization, accusations of unfair treatment of the indigenous community, and disrespect for the present culture. This has been manifested by street protests, attempts to change the date of the National Day and vandalism of national monuments (Sanda, 2020). It is true that, as in most countries, Australian history includes examples of injustices, racism, and bigotry. Although it is right to point out these mistakes, it would also be fair to acknowledge the great accomplishments that the nation has achieved. An imbalance in teaching history can result in disharmony and lack of national pride and patriotism, essential qualities for cohesion in a society.

Questions to Ponder

I have tried to highlight what I believe are some of the mistakes we have been making, emphasizing two in particular, the response to perceived

AGW and the response to the coronavirus pandemic. To conclude, I will pose two questions that relate to these issues:

1. How do false beliefs come to be widely accepted by a large proportion of the public?
2. How do politicians decide to introduce policies based on these false beliefs that can cause serious damage?

Of course, some may argue that what I assert to be false beliefs are not false and this is a legitimate stance. It is how the scientific method should operate, by a contest of different opinions. In the next chapter, I will try to defend my position by suggesting how false ideas and policies have become accepted. In keeping with the title of the chapter, however, I suggest that the mistake that has been made is that decisions have been adopted without sufficient public debate.

References

Baggs, M. 2019. BBC News, Gillette faces backlash and boycott over "MeToo" ad. January 15.

Bagus, P., Pena-Ramos, J.A., and Sanchez-Bayon, A. 2021. COVID-19 and the public economy of mass hysteria. International Journal of Environmental Research and Public Health 18: 1376.

BHP 2020. BHP climate change briefing, BHP briefing webcast. 10 September.

Capuzzo, M. 2021. The drug that cracked COVID. The Magazine of Pennsylvania & the New York Finger Lakes, Mountain Home.

Clayton, R., 2020. ANZ's climate policy steps away from coal to support net zero emissions by 2050. ABC News, posted October 29, 2020, updated October 30, 2020.

DOCS4 Open Debate 2020. Open letter from medical doctors and health professionals to all Belgian authorities and all Belgian Media. September 5.

Fox, C. 2016. Generation Snowflake: how we train our kids to be censorious crybabies. The Spectator, London, June 4.

Henderson, G. 2020. Disagree with other views? Shut down the platform. The Australian, November 21.

Hinderaker, J. 2019. The environmental disaster of solar energy. Center of the American Experiment, August 15.

Investor's Business Daily. 2018. Media Bias: Pretty Much All Of Journalism Now Leans Left, Study Shows (Editorial), November 16.

Kelly, J. 2019. Climate change debunked. Which end is up? Tools for Disorienting Times, October 1, 2019. www.whichend.com/post/climate-change-debunked

Popper, K.R. 2002. Conjectures and refutation. Routledge, London.

Qantas 2019. Qantas confirms commitment to social justice issues, including LGBTI, indigenous and women's rights. Business and Human Rights Resources Centre, May 9.

Robinson, C. 2016. Our young people are being indoctrinated towards a left wing bias. Blog June 30, 2016.

Sanda, D. 2020. Arrests as more Australian monuments defaced. The Australian, June 14.

Sloane, J. 2020. The season's bumper crop from Left's Long March. The Australian, November 14.

Trump, D. 2020. Remarks by President Trump at the White House conference on American History. National Archives Museum, Washington, DC.

Tybalt, M. 2017. Left-wing indoctrination is in full swing and not just in the way you think. The Unshackled. Breaking the chains of control. July 17.

chapter eight

How Do We Fix What Has Been Going Wrong?

A traveller on a trip to a rural area stopped to ask an old-timer for directions on how to arrive at his destination. The old-timer answered, "If I wanted to go there, I wouldn't start from here". If we want to fix some problems that we perceive to exist, the advice might be similar. It might be better to go back in time before some present policies were put in place. Unfortunately, we don't have that option. Mistakes have been made and mistaken policies adopted. It is not possible to put our gears in reverse. We are stuck with having to start from where we are.

The first aim is to decide where we are going wrong and some of the mistakes I believe we are making were described in Chapter 7. If we can recognize mistakes, at least we can plan strategies to correct them. This will enable us to see more clearly the path needed to approach our destination. As hinted to in earlier chapters, our destination is a society made up of free thinkers and the absence of dogmatic ideologues and their accompanying followers. The path must be taken with courage but also humility. We need to admit that we don't have all the answers but we do have a criterion for recognizing mistakes. If we use critical thinking and apply the scientific method, we are able to learn from our mistakes and to change course and search for better alternatives.

Where do we start? It has to start in the school system. That is where I believe we have started to go wrong. Children are malleable as was astutely recognized by Vladimir Lenin. If certain seeds of belief are sown in their minds at this early stage, they may not be able to free themselves from these beliefs as they go through life. The children of today will be future generations of adults, becoming, in time, leaders who influence the direction a culture takes. In the 20th century, extreme ideologies flourished, both extreme left (communism) and extreme right (Nazism). Western countries are now largely, although, not completely, free of these influences. Both are examples of how extreme ideologies lead to control of citizens and the suppression of liberty and creative thinking that have produced the great achievements of the past.

The previous chapter described two contrasting paths that a society can proceed along. One leads to maximum freedom for individual citizens consistent with not impinging on the freedoms of others. The other is one

DOI: 10.1201/9781003254065-8

that suppresses freedom of individuals in a quest for the best interest of a collectivist society. Earlier chapters have described some of the mistaken views that may be negatively impacting our way of life. We will now focus more closely on identifying them and on how we might take action to correct them. It is not intended to assume that we know the truth. Many may disagree with what is claimed. They may say that mistakes are not being made or, if they are, the proposed solutions are wrong. In scientific debate, criticism is welcome. The rational discussion that follows can often lead to a greater understanding for all.

Before discussing how to address the alleged mistakes, it might be useful to say something on the subject of wokeness that I think is relevant and serves as a prelude to the questions we are going to consider. There is no clear definition of the word woke but its meaning is generally understood. Some of what it encompasses has been referred to in Chapter 7 in regard to an article by Judith Sloane. Woke topics included the evils of capitalism, colonialism, racism, women's rights, inequality, diversity, multiculturalism, intersectionality, and climate change, all of which come under the broad umbrellas of political correctness and social justice. Wokeness has infiltrated our culture in recent times. In a similar way, ideas have been introduced in the past and supported by large numbers of followers. A small group of elitist authoritarians are able to impose their ideology with the help of a cohort which Vladimir Lenin dubbed "useful idiots". It included people who are unable to think critically and are brainwashed into adopting certain ways of thinking. If wokeness is enabled to develop, it will inevitably lead to an illiberal movement that resembles communism or fascism. We are already seeing signs of this approaching with the shutting down of free speech and the formation of a cancel culture.

We always need to be vigilant to prevent our culture from being taken over by some dangerous ideology. To ensure this, all we need to do is replace the useful idiots in our society by free thinkers who have the capability to critically evaluate different ideas and refuse to be indoctrinated. There's the answer. Easy! Done! Problem solved. Who would have thought that the solution would be so easy? But hang on, can it be that easy? I will return to this question a bit later.

Resilience

The topic of resilience was introduced in Chapter 1 and expanded on in Chapter 7, referring to an article by Claire Fox (Fox, 2016). How do we nurture greater resilience in our society? Obviously, we need to avoid the pampering pointed out by Claire Fox and present young citizens with the challenges that are needed to build fortitude. One way to do this is to reintroduce citizen military training. The phasing out of school cadets and National Service training in Australia may have been mistakes although probably they have been made with the best intentions. Mistakes are acceptable providing they are recognized and followed by attempts to correct them. If we eliminate any citizen military training, that shows that we are not a belligerent nation and that therefore we are virtuous. This would work well if everyone else in the world shared the same high ideals. History has shown us that this is usually not the case.

When I began, I had the intention to keep the book apolitical and I have tried to do that. At this stage, I feel obliged to stress a point. The criticisms of the extreme political left are not based on political bias. They are based on the scientific principle that we should recognize mistakes, learn from them, and advance from there. If the mistakes had come from the political right (for example, if they were made by fascists), I would have been equally critical. The principle is that a movement towards extremes is a move toward totalitarianism and this, to my way of thinking, is to proceed along a mistaken path.

In Chapter 4, we looked at some criteria for identifying warnings of a move toward extremism. These included a refusal to accept the result of a democratic election, attempts to silence opposition, lack of humour, and the distortion of facts. One of the problems for those who move too far towards an extreme political view is that they become unable to admit mistakes. They have built their beliefs on a certain platform. If they proceed too far along that path, a point of no return is reached. They have invested so much capital that it becomes impossible to reverse.

How do we detain the indoctrination of students in schools? Young people, as already mentioned, are impressionable and can be easily influenced. To counter the manipulation of young minds, the most effective way is to provide instruction on critical thinking. It is a topic that has not been given sufficient emphasis in school curricula. It has been briefly mentioned in Chapter 3. Once a person becomes proficient in critical thinking, that person is less vulnerable to mental persuasion. The goal of instruction in critical thinking is to produce students who are capable of evaluating arguments objectively and not swallowing everything they hear or are told. If they acquire this capacity and learn not to adopt fixed viewpoints based on limited information, then they will have a defence against brainwashing. It acts like inoculation against a disease. The question is how this instruction should be imparted. The best way may be not to merely rely on teachers, but to contract experts who are qualified in teaching critical thinking and will give instruction free of political bias. Rather than entirely depending on schools to provide the teaching, a more appropriate method would be to utilize the internet to enable presentations by experts. The class teacher could then follow up by inviting feedback and discussion from the students. The presentations from the experts would of course need to be closely monitored by independent observers to ensure the absence of bias. More extensive use of the internet, which is already happening slowly, to provide students with access to information from the best minds, would be valuable for all subjects.

In addition to instruction on critical thinking, two other complementary areas of instruction that should be included in schools are the philosophy of the scientific method and political ideology. Both these topics have been covered in earlier chapters. The scientific method, in particular the hypothetico-deductive method, was described in Chapters 2 and 3. A sensible method of teaching political ideology is described in Chapter 4. The approach used by Kate Habgood is one that encourages students to think about political issues and form their own opinions without outside influences, i.e. free of brainwashing.

Let us now focus on specific mistakes that have been made and explore how they might be corrected. In Chapter 7, areas were listed where I believe we are going wrong. If we can admit that we have made mistakes, this opens up opportunities. We can learn from the mistakes and try alternative ways to proceed. This is the scientific method. It is the method that has been successful in the past and has led to great advances for humanity. The alternative is to not recognize that we have made mistakes and are sure that we are following the right path. Thus, we have a fork in the road with two alternative paths. Which one do we choose? We can take one fork where we reflect critically on our ideas, reject those that are shown to be mistaken, and think afresh how we should proceed. The other path is one where we are unable to recognize that we could be wrong and push on, believing that we know the truth. The latter path is fraught with danger. If it turns out that our beliefs are mistaken, we will go through life without ever knowing that we are going wrong.

In Chapter 7, we selected two issues on which to concentrate our focus. These were anthropogenic global warming (AGW) and the imposition of lockdowns in response to the coronavirus pandemic. The two issues were chosen because they share common features. First, they are both promoted by creating fear in the population. Second, both have led to policies that are causing serious damage to many individual's lifestyles as well as destroying businesses and country's economies. Third, the responses to the two issues are touted as being based on science. I believe that the first and second common features are true but I will argue the case that the third is false.

Anthropogenic Global Warming

The belief that human activities are making a significant contribution to warming of the planet has been largely influenced by the regular reports of the United Nations (UN)'s Intergovernmental Panel on Climate Change (IPCC) with the help from a compliant media. These have been instrumental in persuading the general public that the prognostications of the

IPCC are based on science and therefore should be accepted. In Chapter 7, arguments were presented to refute the assumption that the work of the IPCC is scientifically sound. However, the position of the IPCC which has been to forecast dangerous future warming of the planet, unless mitigating policies are introduced, has become the politically correct viewpoint. It has been uncritically adopted by many of those in power, resulting in policies to phase out fossil fuels and to subsidize intermittent energy sources. These policies are having and will have destructive effects on the economies of some countries.

Now that the acceptance of anthropogenic warming has been inculcated in a large section of the public, there is a battle in progress to try to convince people to rethink the issue. To acquaint the reader with the nature of this battle, I have selected two campaigns in which opposing viewpoints are presented. The first one is based on an article entitled "World Scientists Warning of a Climate Emergency" (Ripple et al., 2020). It was put forward by an entity called "The Alliance of World Scientists" and is a declaration which claims to have more than 11,000 scientist signatories from some 150 countries around the world. It states "clearly and unequivocally that planet Earth is facing a climate emergency". The steps needed to lessen the worst effects of climate change are listed. They include replacing fossil fuels with renewables and leaving stocks of fossil fuels in the ground. An extension of this aim is that wealthier countries need to support poorer countries in transitioning away from fossil fuels.

A humorous critique of the declaration was posted on YouTube entitled "11,000 scientists warn of untold suffering from climate change. Meet the Canadians who made the list" (Ezra Levant, Rebel News, November 7, 2019). In this presentation, Ezra Levant traces the "Alliance of World Scientists" to a short blog on a home-made website. He exposes how many of the signatories were not scientists and were people who had clicked on the web page in a similar way to how someone clicks "like" on a Facebook page.

In sharp contrast, there is a website in which over 900 individuals, mostly genuine scientists, have signed a declaration that there is no climate emergency. They belong to a foundation called Climate Intelligence (CLINTEL) which was formed in 2019. The names and credentials of the signatories are listed on the CLINTEL website (www.clintel.org). They have been carefully vetted to confirm that they are genuine and are qualified to hold opinions on the subject of climate change. The message put forward by CLINTEL is that climate science should be less political and climate policies should be more scientific. It states that climate science has degenerated into a discussion based on beliefs, not on sound self-critical science. We should free ourselves from immature climate models. What comes out of climate models is only dependent on what theoreticians and programmers have put into them. The main objective of CLINTEL is to generate knowledge and understanding of the causes and effects of climate change, the policies that have resulted and to expose where facts turn into assumptions and predictions. It would like to stimulate public debate and to function as an international meeting place for scientists with different views.

What can we take from examination of these two divergent declarations? One possible conclusion is that, since 11,000 is a much greater number than 900, this must carry the greater weight. This is where critical thinking needs to be applied. A closer scrutiny reveals that the signatories to the "World Alliance of Scientists" petition simply resulted from many who were not true scientists and clicked on the equivalent of a "like" button on a Facebook page. The signatories for the CLINTEL declaration, on the other hand, are all people who have been closely vetted and are qualified to express an opinion on the issue.

The belief that there are substantial human effects on climate has been accepted by a large section of the community. The justification for the belief is widely attributed to "The Science". Let us look more closely at what is meant by "The Science". The body most influential in consolidating the public's view is the UN's IPCC. This is a large well-funded organization that comprises three working groups and a task force. Group I deals with the scientific basis of climate change, Group II with climate change impacts, and Group III with mitigation of climate change. Each group extracts information from a large number of scientists who work in the area of climate science. They then summarize the data to produce regular reports which are accessible to all who are interested. The work of the IPCC appears to be carried out diligently, responsibly, and honestly. There are criticisms, however, that need to be brought up and these are as follows:

1. The aims of the IPCC are based on the premise that human activities (sending greenhouse gases into the atmosphere) are making a significant contribution to warming of the planet. Scientific investigators should never begin with assumptions. They need to have no preconceived beliefs and be open-minded to any possibility. There has been no statement as to what is the ideal average temperature of the planet. It could be lower or it could be higher than the present value. In general, higher temperatures lead to lower human mortality and to greater greening, resulting in more abundant harvests. What is the rationale for believing that warming is detrimental? Working group III aims at "tackling" climate change by finding ways to decrease the amount of greenhouse gases in the atmosphere without having established that this would be beneficial.

2. The modus operandi is to aim at reaching a consensus among scientists. Consensus is not a criterion for the validity of a theory. Frequently, scientists whose views differ from the mainstream are those who contribute to genuine advances in knowledge. If the views of a minority are excluded from the IPCC reports, it may well be that

the minority includes some who hold the more correct opinions. Just because a majority share the same viewpoint, this does not justify accepting it as the correct one.

3. The process used to gather information with the view to arriving at a generalization is the inductive method. Induction is not a valid method for advancing knowledge as has been pointed out by Karl Popper, one of the great philosophers of science. The correct method is that of hypothesis-deduction.

4. If we take as an example the report of working Group III, we find terms such as "virtually certain", "highly likely". These are not scientific terms. In the hypothetico-deductive method, a conjecture (hypothesis) is proposed to explain observations or answer a question. It is then subjected to severe tests. The tests may refute the hypothesis, they may corroborate it or they may be inconclusive. There is no scope for terms such as "highly likely".

The most valid argument to question the deliberations of the IPCC is that given in Chapter 7. The procedure of the IPCC in seeking information from a large number of scientists who are active in the area of climate science and then attempting to arrive at a consensus is not the true scientific method. The correct procedure is to allow individual scientists to debate the issue and, through a dialectic approach, to get closer to the truth. It is surprising that more scientists have not come forward to expose the fallacy of the IPCC approach.

When an organization is formed with a stated purpose, activists can often infiltrate and use the organization to assist in imposing additional agendas different to the initial one. This seems to have happened with the UN's IPCC. At the 52nd session of the IPCC in February 2020, the panel adopted the Gender Policy and Implementation Plan. This was built on a report from the IPCC task group on gender at their meeting in Kyoto on May 8–12, 2019. As part of the mandate of the Implementation Plan, it was stated that activities will be undertaken to enhance gender equality and provide training and guidance on gender-related issues. This is far from the original aim of the IPCC which was to focus on climate issues. The adoption of policies not related to climate casts doubt on the authenticity of the IPCC. It seems to have become more about taking advantage of UN funding for junkets to advance woke agendas.

Lockdowns in Response to the COVID-19 Pandemic

There has been much written about the COVID-19 pandemic and how different countries have dealt with it. As stated in Chapter 7, one of the most plausible evaluations of the responses has been presented in an open letter from Belgian doctors and medical specialists. The thrust of the letter is that policies introduced by governments, such as lockdowns of citizens, are not sufficiently scientifically based. Another criticism is that there has been insufficient opportunity for an open debate in the media. The result of these deficiencies has been the imposition of regulations that have been, to a large extent, left in the hands of small numbers of bureaucrats. It has resulted in the promotion of arbitrary policies that have been inconsistent between different jurisdictions.

The true scientific position is always that we need to be open to opposing viewpoints. The imposition of strict authoritarian lockdowns in several countries has resulted in the reduction of new infections to close to zero. This has been the case in China, Singapore, New Zealand, and the state of Victoria in Australia. Time is needed to adequately assess the efficacy of lockdowns. There are questions that need to be considered as the future unfolds. Focusing solely on the reduction of infections is only part of the big picture. There are other effects to consider. What has been the collateral damage resulting from lockdowns? What will be the effect on natural immunity in future pandemics? What are other health aspects such as mental and psychological damage and failure to show up for routine medical tests that could have serious consequences?

Policies adopted to address anthropogenic climate change (ACC) and those imposed to counter spread of the COVID-19 virus have features in common as mentioned previously. Both claim to be based on science but neither has a solid scientific basis. Policies that have been introduced to supposedly address the two issues are causing hardship to individuals and businesses as well as damage to the economies of some countries. If they are mistakes, what has to be done to correct them? In the case of the climate issue, CLINTEL, possibly the most authentic association of scientists who are qualified to deliberate on the issue, submits that there is no climate emergency. Therefore, no actions are needed and it follows that policies such as phasing out of fossil fuels and their substitution by intermittent energy sources should be discontinued. In the case of the COVID-19 pandemic, the open letter from the medical and health specialists (initially from Belgium) could be the most authentic appraisal of the policies that has been introduced. Like the climate specialists, this group of knowledgeable medical and health experts call for an immediate end to all the measures that are being imposed.

Is it likely that these recommendations will be accepted? Unfortunately, it is not. There are enormous vested interests that will resist any moves to dismantle the measures that have been put in place. There is no need to invoke a conspiracy plot although this may well play some part. There are those who profit from continuation of the measures. I will give a few examples but there are many. Those who manufacture wind turbines and solar panels would naturally resist having these industries terminated, particularly as they are assisted by subsidies. There are those who are ideologically opposed to political systems such as exist in Western democracies. Any damage that can be done to those systems would meet with their approval. Another factor is that there are individuals who love to be in a position of power. They may use the power to control citizens or help to win elections. In a lockdown situation, people can develop Stockholm syndrome. This is where hostages feel affection and trust for their captors. In the case of the COVID-19 lockdowns, the rate of infection often drops over time. Frequently, this may be just fortuitous but can induce a feeling of gratitude in the captives to those who have ordered the lockdowns. In the Australian states of Queensland and Victoria where some of the strictest controls on freedoms have been imposed, governments have been rewarded by strong support in the electorates.

However, perhaps the greatest obstacle but the one that at least is possible to overcome is the lack of knowledge among the general public. Every one of us is included in this cohort. We all can be hoodwinked because we are unaware of everything that is happening, sometimes clandestinely. Most people are fully occupied. They have to go to work, they have to earn money to put food on the table, to educate their children, to pay their mortgages and health bills. We don't have the spare time needed to fully comprehend what is happening around us. Some of what is happening is having harmful effects on our lives but we are not aware of it. The other fault we all suffer from is our gullibility. We tend to believe what we hear and read in the media. Much of it is true but much is false or at least distorted. There is also information that is withheld or difficult to access. I come back to a point that is repeatedly made in the book. We need to develop our critical thinking skills so that we will not be duped and become "useful idiots". We will not readily accept what is fed to us. We will be able to recognize mistakes, including our own. This will give us power and not allow ourselves to be deceived.

Our current lack of awareness is playing into the hands of those who exploit it to further their own agendas. Even many who could influence policies, such as our politicians, are ignorant of what is going on. They don't have a good understanding of science and bend towards what appears to be public opinion. Acceptance that ACC is a serious problem has become entrenched and is now difficult to counter. Little notice is being taken of dissenting views from authentic bodies such as CLINTEL. It needs input from people who can wield influence. Fortunately, there are a few of these although nowhere near enough. One who is cognizant of the true situation is a federal senator from a minor Australian political party, One Nation. Malcolm Roberts has a background in mining engineering and has expertise in atmospheric gases. He has been a lone voice in putting forward arguments to the Australian parliament that there has been no reliable scientific evidence that human activities are contributing to climate change.

The Australian government supposedly takes its scientific advice from its premier research organization, the CSIRO. The leaders of this organization seem to have adopted the view that ACC has been established. Senator Roberts has been scathing in his criticism of CSIRO. From discussions with their staff, the senator concluded, in a report, that CSIRO (1) has relied on discredited and poor quality papers for "evidence", (2) has never stated that today's temperatures are unprecedented, (3) has never quantified any specific impact of CO_2 from human activities, and (4) has allowed politicians and journalists to misrepresent CSIRO science without correction. Senator Roberts, in his report, recommends that "Until the government provides scientific proof of specified quantitative effects of human CO_2, all climate policies need to stop immediately". Another recommendation states, "The parliamentary debate that has never been held, needs to start with parties that are advocating climate policies presenting to parliament their empirical evidence in a framework proving causation and justifying their climate policies with specific quantified targets and impacts". In view of the huge economic effects of the climate policies that are being imposed around the world, this sounds like a perfectly rational request.

School Education

The excellence that a society attains is to some extent dependent on the quality of education given to children in schools. Some criticisms have been made of the Australian school curriculum which was put in place in 2015. One of these was the incorporation of cross-curriculum priorities that were based on political ideology. A new curriculum is currently being developed that is planned to be in place by 2022. It is hoped that this will give the opportunity for mistakes in the present curriculum to be corrected. I will therefore not resort to further criticisms and will await with hope for a new and improved curriculum.

Only a few suggestions will be made. Education is much more than simply teaching the traditional disciplines. It should also incorporate policies to improve the well-being of young people and fit them to become valuable citizens. We need to pay attention to the ideas of John Gotto and John Medina which were covered in Chapter 6. Application of recent research in the area of positive psychology should be included in the curriculum. The ideas of psychologists such as Martin Seligman (Chapter 6) would be beneficial. Helpful advice on building resilience and feeling gratitude would be antidotes to some of the negative behaviours that can occur such as depression and drug addiction.

One of the problems associated with the lowering of academic performance noted in Chapter 3 was the disruption caused by some students. It has become increasingly difficult for teachers to deal with this behaviour. In recent times, there has been a removal of the disciplinary methods that teachers are permitted to use to bring disruptive students into line. In order to prevent further erosion of order in classrooms, some innovative policies are needed. In order to capture the attention of students so as to reduce the distraction that some cause, it would be better, rather than meeting out punishment, to make the classes more interesting. One example of how to achieve this is teaching of magic. Magic includes many fascinating events such as making things vanish or predicting the future. The ability to perform magic tricks requires considerable practice. It could be a valuable part of teacher training. It is just one innovation that could counter the lack of interest that leads to disruptive behaviour. Many useful links to "teaching with magic" can be found on the internet. The contribution of Kevin Ogren is one example (http://hdl.handle.net/1828/5280). Here, Ogren states, "Magic is an effective strategy to motivate and inspire students to read, advance their physiotherapy, build confidence and think creatively".

Tertiary Education

Several problems in tertiary education have manifested in recent times. Making it available to all (initially by making it free and subsequently through loans) has produced an excess of graduates in some areas. Many who graduate are unable to find employment in their chosen field. Dependence on foreign student fees to make universities viable has created a fragile system that is vulnerable to failure. There is also a question as to whether academic standards are being maintained. Some of the foreign students do not have a strong command of the English language. One of the incentives for enrolment of foreign students is to enhance their chances of acquiring permanent residence on completion of their degrees. The reluctance of universities to welcome financial donations to support courses on Western Civilization is worrying. The rejection appears to emanate from the extreme left-wing culture that has encroached into universities, particularly in the humanities departments. The acceptance by three universities is a welcome development. It should provide a bulwark to leftist indoctrination with its accompanying denigration of Western culture. There may be arguments for introducing stricter selection criteria for admission and for limiting the number of foreign students being admitted.

Corporations

Movements towards the political left are understandable in the education field and the media. What appears puzzling to many is how some leaders of business enterprises have risen to power where they adopt woke policies. Normally, it would be expected that CEOs of these corporations would be more on the conservative side to be consistent with capitalist ideals. It seems paradoxical that some of the appointees in recent times foster policies that appear opposite to those aimed at enhancing the prosperity of companies. They justify their position by arguing that embracing social justice issues is not incompatible with the main aim of a business, which is to maximize profits. One example is the policy of some companies to support the target of zero industrial emissions by a future date. This can lead to financial losses. The justification for the policy is that the financial losses that occur could be less than those that might occur by ignoring the effects of future climate changes and the regulations introduced to combat them. It is based on the acceptance that serious weather events (higher temperatures, increased fire danger, cyclones, and floods) are likely to occur more frequently in the future. A critical examination shows, however, that severe weather events are not happening more frequently.

How do we understand these decisions made by corporate leaders? It shows that those who have pushed the theory of ACC have stolen the march over the true scientists (e.g. the members of CLINTEL). It will not be easy to overturn their influence. How could such people have reached the top? Evidently, they do so by adhering to political correctness. Most people in a corporation don't have the knowledge to understand an issue such as ACC. If there are some who do, they will find it hard to challenge the views of the politically correct. It has taken these two issues, ACC and lockdowns to expose the situation where a great majority are persuaded to submit to a small minority. It is shown by the readiness of some to accept the theory of ACC and others who show gratitude to their captors for protecting them from a virus. Perhaps more than ever before, these events draw attention to the need to develop critical thinking in young people so that, when they mature, it will not be so easy to fool them.

The Media

A free media is an important component of a democratic society but, in Chapter 7, we saw how it can be infiltrated by bias. Once bias becomes entrenched in a media outlet, it is difficult to change the culture. Young journalists are pressured to conform to the prevailing ideology. One suggestion on how to resist the trend was made in an article by Chris Mitchell (Mitchell, 2020). The article criticises university journalism education in "creating a class of young reporters more interested in driving social changes than reporting fairly about it". To counter this, he has advised prospective journalists to study other disciplines such as economics, law, or generalist arts rather than journalism. Perhaps a science background might also be a good source for recruitment of journalists. Providing that the training included philosophy of the scientific method, these recruits may help to restore some of the objectivity that seems to have been lost. In a similar vein to Mitchell, Janet Albrechtsen, in an article critical of the Australian public broadcaster (Albrechtsen, 2020), states that

> The ABC recruits from the same pool of woolly-minded idealists with journalism degrees from the same universities. They were cultural clones when it came to climate change, asylum seekers, gay marriage, dislike of religion and preaching their own set of rigid commandments about identity politics.

Bureaucracy

The bureaucracy is there to ensure that regulations are being adhered to and, importantly, to assist the general public with the problems they face. Most bureaucrats perform their tasks with credit. There are always some, however, who give the impression that they believe their job is to find obstacles to put in the paths of citizens. In recent times, this has been evidenced by a minority who wield considerable power by excessively restricting freedoms of citizens during the COVID-19 pandemic. Bureaucrats are expected to be subservient to elected representatives of the people. They should not have the power to sign international, agreements, especially ones that are not necessarily supported by the majority of citizens.

Judiciary

Judges are expected to interpret the laws as they stand. On occasions, some judges can make decisions that reflect their political leanings. It is referred to as legislating from the bench. This practice has to be closely scrutinized and its practitioners held to account.

Concluding Comments

In the previous chapter, I raised two questions:

1. How do false beliefs come to be widely accepted by a large proportion of the population?
2. How do politicians decide to introduce policies based on these false beliefs that can cause serious damage?

In his excellent book "The Parasitic Mind: How infectious ideas are killing common sense", Gad Saad explains how false ideas can be inculcated into a society (Saad, 2020). These ideas spread like viruses. One of the unfortunate consequences is that, once the idea has been accepted, it is extremely difficult to eradicate. For many humans, a belief, once adopted, tends to become entrenched. The antidote for this behaviour is application of the scientific method. That is why, as emphasized repeatedly, teaching of the philosophy of the hypothetico-deductive scientific method is so important for inclusion in education. The true scientific method of thinking never assumes that the truth has been obtained and one always remains open to changing an opinion. This approach thus protects us from adopting entrenched false beliefs.

A question was posed (Chapter 4) as to why there is a division on political grounds about AGW when it is purely a scientific issue. One possible answer is that it could be the result of the contagious nature of ideas as described by Saad. Those on the political left predominantly support the theory while those on the right tend to be sceptical. People tend to associate with others who hold the same beliefs and listen to views expressed in the media which share their opinions. The UN, particularly its IPCC, supports the theory. The UN is a body that veers to the left and is becoming increasingly so as it stacks its leadership with people who support policies such as World Government. The Mainstream Media (MSM) has, as we have seen, become predominantly left-leaning.

One unfortunate trend has been to erroneously attribute beliefs (such as dangerous AGW) to "the science". An example is the claim that the IPCC reports represent "the science" and therefore should not be questioned. One writer who is not a scientist but is given appreciable media coverage has referred to the IPCC reports as the "gold standard" for climate science. Earlier in the chapter, I have pointed out that summarizing peer-reviewed scientific literature to arrive at a consensus is not true science. This is apart from the fact that much of the peer-reviewed literature on climate is based on unsubstantiated modelling rather than empirical testing.

References

Albrechtsen, J. 2020. Ita Buttrose needs to show respect by changing "smug and boring" ABC. The Australian, December 16, 2020.

Fox, C. 2016. Generation snowflake: how we train our kids to be censorious crybabies. The Spectator, London, June 4.

Mitchell, C. 2020. Left's lone march, social media and educational theory murder journalism. The Australian, November 22.

Ripple, W.J., et al. 2020. World scientists' warning of a climate emergency. BioScience 70: 8–12.

Saad, G. 2020. The parasitic mind: how infectious ideas are killing common sense. Regnery Publishing, Washington, DC.

chapter nine

Summary and Hopes for the Future

In its simplest form, the scientific method can be thought of as learning from our mistakes and trying to correct them. True scientists try to think rationally, never adopt dogmatic opinions, and are always willing to listen to opposing views. They never claim to know the absolute truth but are relentless in their search for it. Let us now summarize some of the issues that we have been considering. Before doing so, let me reiterate: The views I express are ones that I currently hold. They are not rigid and can be altered by considering new facts or rational arguments.

1. We are not providing the conditions for young people to develop resilience. One way of addressing this is to resume national military service. This would have two positive outcomes. It would help develop resilience of individuals and it would enhance security of the nation in times that are becoming increasingly uncertain.
2. We are allowing young people to be subjected to indoctrination by ideology that could be harmful. To prevent this undesirable activity from escalating, there needs to be increased attention given to instruction on critical thinking in education. This will help to produce citizens who are free thinkers and are not vulnerable to easy manipulation. The lack of critical thinking has been clearly exposed in the recent pandemic in which large numbers of people have happily accepted their loss of freedoms and shown gratitude to those who have essentially acted as their captors by imposing arbitrary restrictions.
3. Ideological beliefs should not be incorporated into school curricula as has been done with the imposition of cross-curricular topics. The effects of indoctrination and groupthink are impacting negatively on different areas of society and particularly on the media. Doctrinal beliefs instilled into reporters are passed on to the public who then may become influenced.
4. Development of critical thinking in education should be extended to an understanding of the hypothetico-deductive scientific method and how it can be applied to everyday issues. Mistaken beliefs such as acceptance of consensus as a criterion for arriving at truth and

DOI: 10.1201/9781003254065-9

equating correlation to cause-effect should be clearly explained once the true scientific method has been outlined. The acceptance of these errors in thinking to arrive at erroneous policies needs to be illustrated by examples such as the climate change controversy.

5. The good points of left and right political thought should be explained and students allowed to form their own opinions about issues, free of coercion.

6. The current teaching of history in schools should be revised to lessen bias. There is a need to present a fairer balance between the mistakes and the achievements that have been made. This will help to create more national pride and patriotism.

7. In forming school curricula, examination of the ideas of John Medina to foster more physical activity and John Gotto to encourage greater curiosity should be among many issues that need to be constantly examined. Teaching of positive psychology concepts according to Martin Seligman and others should be incorporated. This would help to counter the depression and negativity that seem to be increasing.

8. The decision to make tertiary education open to all has had some deleterious effects. For example, there has been an oversupply of graduates in certain fields so that many of them are unable to find jobs that fit their qualifications. Many who have been given the opportunity to take out loans for their studies and have not completed their degrees are leaving the government (i.e. the taxpayer) with ever surmounting debt. There is a need to return the focus of university education to meritocracy.

9. The belief that human activities are making a large contribution to climate change needs to be challenged. The falsehoods that are being promulgated need to be exposed and the opinions of true scientists should be given more exposure. A group of eminent scientists, the Argonauts, have shown that the original estimates of global warming have overestimated the effect by a factor of three. When this mistake is corrected, the predicted warming agrees more closely with what is being observed. It also means that the estimated global warming is not harmful. There needs to be more open public debate on the issue, not talkfests in which the agendas have been set beforehand. Until it is shown by TRUE science that emissions from fossil fuels are contributing to harmful effects on the climate, the policies being adopted to phase out these efficient energy sources should be terminated immediately.

10. Energy policies based on the phasing out of fossil fuels and substituting them with intermittent energy sources need to be questioned and scrapped if they are shown to be not justifiable. The development of nuclear energy should be given high priority. It has been shown to

be safe and is used successfully in countries such as France. Modern modular nuclear reactors appear attractive for the future.

11. Following considerations 9 and 10, it needs to be recognized that Australia is rich in resources for supplying fossil fuel (coal) and nuclear fuel (uranium). These resources should be exploited for domestic use. It is reprehensible that they are not and the population as a consequence is being subjected to ever increasing electricity costs.

12. The imposition of lockdowns to counter pandemics needs to be critically evaluated and policies based on them discarded if they are shown to be counterproductive. The damaging effects of mass hysteria need to be investigated and the topic should be publically debated and included in educational courses on critical thinking.

13. Bias in the media seems to be increasing in recent times and the dangers from group think have become evident. Journalists are being sourced from the same university departments of journalism. This is turning out journalists with narrow perspectives. There is a perception among experienced media specialists that the sources for recruitment of journalists should be broadened.

14. It is becoming apparent that some who are being installed as corporate leaders are increasingly adopting politically correct (woke) policies rather than giving priority to policies directed to enhancing the prosperity of the company. Shareholders have the responsibility to be aware of this failure and to rectify it.

15. There are issues that are having huge effects on nations but the general public is not appropriately informed (e.g. anthropogenic climate change, responses to pandemics). Many media outlets project only partisan views. There is a need for the media to step up and organize debates with a neutral adjudicator in which all sides of an issue are presented. This will help to create a public that is better informed than currently.

16. There is a need for statesmen/women to step up. These are citizens who give priority to the best interests of the country, not consideration of how best to get elected. Even if such individuals fail to be elected, by voicing their views, they will influence public opinion and guide it towards more noble goals.

chapter ten

What Are Some of the Mistakes and Can They Be Corrected?

We don't know how our future will turn out. This is a good thing because, if we accept that the future is not predetermined, this means that we are in a position to take actions that will give us the best outcome. What are the mistakes that have been made? If we recognize them, then we can correct them to enable us to make things better. Let us proceed on the basis of this simple idea. It would be too ambitious to embark on a detailed list of mistakes that have been/are being made. What will be attempted is to try to point out a few and how they might be corrected.

Scientific Community

As a member myself, I accept that the scientific community has been weak in defending scientific principles and has allowed false beliefs to be inculcated in our society. These false beliefs have led to what I believe are mistaken policies that are having disastrous effects. There have been some scientists who have stood up but there has not been anywhere near enough and so their voices have been drowned out by activists who have been successful in indoctrinating large sections of the polity, the bureaucracy, and the general public. This has been facilitated by the fact that most of the public is ignorant of the true scientific method. The philosophy of the true method which is the hypothetico-deductive method should be taught from an early age. The teaching should include instruction on the limitations of the inductive method and the false conclusion that correlation equates to cause-effect.

The Teaching and Education Community

School children have been indoctrinated by political ideology. This is creating a cohort of citizens who have been brainwashed to believe specific political doctrines such as those that come under the heading of wokeism. It begins at primary and secondary levels of schooling but has advanced to infiltrate tertiary institutions. We can blame the teachers but we need to examine how they have been able to get away with it. When this is done, we see that there are many who share the blame for the unfortunate trend.

DOI: 10.1201/9781003254065-10

It includes members of education departments, the teachers' unions who have been shown to support policies that lead to lowering of educational standards, and, above all, the politicians who have permitted it. There is an urgent need for the mistakes to be recognized and policies put in place to correct them. This includes reinforcing teaching of critical thinking, teaching of the true scientific method and instruction on political ideology free of brainwashing. Education in schools should include guiding students towards organizations that provide objective information such as the Institute of Public Affairs (IPA) in Australia and away from the indoctrination and groupthink that they can find on social media sites such as Twitter.

The Psychological and Psychiatric Community

As noted in Chapter 3, the effects of bullying and other antisocial behaviour have had negative influences on schoolchildren and their performance. This has extended into the workforce. There have been individuals who have acquired the highest leadership positions in the country who are psychopaths. Very few in the general public understand this personality trait which can have destructive effects out of proportion to its frequency. Those who should understand (psychologists and psychiatrists) have been weak in stepping up to educate the public to their dangers and to identify the transgressors. This cohort of specialists needs to show more courage and strive to fulfil their responsibilities to the community.

Politicians

Our elected representatives are involved in all the areas we are examining. They are responsible for maintaining educational standards. They have allowed pernicious dogma to infiltrate school systems where students have been brainwashed by dangerous ideologies such as catastrophic climate change, gender fluidity, victimhood, and other woke topics. It is they who are at least partially responsible for allowing this negative culture to promulgate. They also, in part, bear responsibility for the decisions made by bureaucrats under their control. Decisions have been made by individuals who have been given excessive power to make arbitrary rules that curtail the freedoms of members of the public. These bureaucrats are permitted to destroy the lives of many in the private sector while they, themselves, keep their jobs and salaries. Politicians have shirked their duty by claiming to rely on decisions made by individual "experts". The policies based on the irrational belief in catastrophic climate change as well as the acceptance of actions by "health experts" to place unjustified controls on citizens must ultimately be sheeted home to those for whom the buck stops, the elected representatives.

The Bureaucracy

Bureaucrats are charged with the responsibility to ensure that regulations are followed and to assist the public in complying with them. Many perform their duties responsibly. However, there are those who abuse their privileged positions. This has been alluded to in discussing the arbitrary, inconsistent, and, in some cases, cruel decisions made by "health experts" to supposedly counter the COVID epidemic. Bureaucrats (health officials) have been given excessive power to impose regulations (e.g. lockdowns, compulsory wearing of masks, and restrictions on travel) that have not been consistent between different jurisdictions such as Australian states. They have been allowed to impose rules that have destroyed businesses, peoples' livelihoods and freedoms. Another local (Australian) failure has been the heartless treatment of military personnel who have returned from conflict (Iraq and Afghanistan). These veterans as well as their families have not been given the support that was merited, resulting in considerable mental problems and suicide (Jones, 2020).

The Judiciary

The decisions of several courts in relation to the prosecution of Cardinal George Pell have been summarized in Chapter 3. The unanimous judgement of the Australian High Court acknowledged that two of the judges in the first appeal had found the complainant to be a credible and reliable witness. However, it pointed to other evidence that was inconsistent with the complainant's account. This evidence should have been considered but was unchallenged. A simple scientific experiment in which the timeline of the complainant's testimony could have been checked was not undertaken at the time. If it had, it would have cast doubt on the accuracy of the complainant's statement and be sufficient to prevent a decision based on "beyond reasonable doubt" as is the requirement. The case with its numerous trials has demonstrated that even our legal system can be subject to mistakes. The final decision of the High Court, however, does reassure us of the reliability of the justice system and for us to have confidence in it.

The Media

Media organizations sometimes show a bias towards one side of the political spectrum. This does not become serious providing there is a diversification of views within the organization. Some media outlets stress the need for diversity but, often, this is meant to refer to such things as sex or race, which are irrelevant. Diversification of opinion is what is relevant. When this is missing, there is no contest of ideas, an important requisite of a sound democratic system. It can reach the stage where the organization recruits only those who subscribe to what is considered to be the correct ideology, thus reinforcing a groupthink culture. In such a culture, individuals lean towards interpreting facts so as to agree with the ideology.

The scientific method provides a guide on how to avoid this mistake. It recognizes that it is easy to find confirmations if one looks for confirmations. It is therefore imperative to resist this temptation by remaining open to possibilities that do not agree with a given belief. This capacity is missing in a culture that slides into groupthink. In Australia, the groupthink culture has become entrenched in the National Broadcaster, the Australian Broadcasting Commission (ABC). This has been pointed out by many but has been especially shown by Andrew Bolt who has a blog site and is also a commentator on Sky News Australia television. He has drawn attention to a number of issues where the ABC has demonstrated groupthink. One of these has been the denouncement of Cardinal Pell for crimes for which he was finally cleared by a High Court decision (7–0) as described in Chapter 4. Another has been support for a storyline by someone who claimed without proof that he had Aboriginal heritage. Bruce Pascoe declared in a book Dark Emu (Pascoe, 2014) that Aboriginal people were not the simple hunter-gatherers that they had been considered to be by most historians but had developed sophisticated agriculture and lived in houses within towns of up to a thousand people. In a recently published book (Sutton and Walshe, 2021), anthropologist Peter Sutton and archaeologist Keryn Walshe, although praising the Aboriginal culture, have debunked the book by Pascoe, arguing that he misinterpreted key evidence about how Aboriginal Australians lived prior to colonization. In a blog, Andrew Bolt has remarked that none of the ABC's staff of some 4,600 had disagreed with Pascoe's version (Bolt, 2021). Pascoe has been rewarded by his admirers by being appointed as Enterprise Professor at the University of Melbourne.

The Public

In a democratic system, public opinion plays a vital role since that should ultimately decide which policies are adopted. Those who are on the side closer to truth can be disadvantaged in many ways. Those who oppose them are more likely to resort to dishonest methods such as lies and propaganda which may convince many who do not have the capacity to think critically. Ideologues who distort truth flourish in authoritarian regimes. These regimes have an advantage over democracies as they can impose their ideologies, exert control over its people, and punish those who do not obey. More democratic systems have less control which allows more division and diversification of opinion. Which system will prevail? Logically, it would be thought that truth should triumph over falsehood. However, if we consider relatively short times, falsehood can often succeed. We should plan to prevent this happening. What are some of the things we can do to prevent it? We must start with education. Political brainwashing needs to be eliminated from school curricula. It needs to be replaced by instruction on critical thinking. This will prevent the formation of "sheeple" and produce a society of discriminating individuals who will not allow themselves to be brainwashed.

Political Systems

In keeping with the book's title, attempts have been made to identify mistakes that have been/are being made in the world. It might be said that it is easy to point out errors. The real question is: can we learn from them and correct them? If we look at history over the relatively short time of the past century, we have seen the emergence of totalitarian regimes that have tried and, in some cases, succeeded in imposing their rule. The main ones have been the German Nazi Party which was defeated in 1945 after a four-year war and the Russian Communist Party that was effectively nullified after a cold war. Other authoritarian regimes have sprung up to impose their control during this time. It has happened in Iran, North Korea, Laos, Vietnam, Cuba, and Venezuela. All these regimes have led to restriction of freedom and death and destruction to large numbers of its citizens. Why has this happened? The simple answer is because their failures have not been recognized. In other words, each regime did not learn from the mistakes that previous regimes had made based on similar ideology. There is a saying that "insanity is doing the same thing over and over again expecting different results".

Democratic systems are not free of flaws. The dangers do not come from tyrannical forces but from the liberty that allows these societies to be infiltrated by false doctrines. False doctrines such as those inherent in wokeism can spread in societies like cancers. What is needed to counter these threats is principled leadership. In Western societies, there is a perception that this has been lacking in recent times. In Australia, the two main political parties have strayed from their traditional bases. The party of the left which has been characterized by support for the working class has moved towards a party of city-based elitists. The party of the right has been infiltrated by individuals who seem to have lost the values on which it was inaugurated by Robert Menzies. Overall, there appears to be an excess of politicians and a deficiency of statesmen/women. In other parts of the world, we are seeing the main traditional parties being displaced.

There are also signs that the two systems of totalitarianism and democracy are moving towards each other. In some cases, totalitarian regimes are adopting the advantages supplied by capitalism. On the other hand, we are seeing crushing of freedoms in what are believed to be democracies. The emergence of the COVID pandemic has provided tools for imposition of lockdowns and use of greater police force to enforce compliance with arbitrary regulations. This is more similar to what happens in totalitarian regimes.

In Australia, there are still patriotic politicians on both sides whose priority is to do the best for their country but there are some who don't appear to prioritize this goal. It may be that there will be a shake-up of the political landscape in the near future. There is a party; One Nation (www.onenation.org.au) is a party that includes parliamentarians who are passionate about putting the interests of the country foremost. Two who hold leadership positions in the party are Pauline Hanson and Mark Latham. A senator in the party, Malcolm Roberts, has been mentioned in Chapter 8. He has been a lone voice in exposing some of the falsehoods being swallowed by the public regarding matters such as anthropogenic global warming. One can only hope that the public can overcome their prejudices, become informed, and support these genuine parliamentarians. Another party that is currently forming is Australia One (www.australiaoneparty.com), led by a former Special Forces lieutenant-colonel, Riccardo Bosi. This party, as described on its website, is "committed to strengthening Australia as a sovereign, self-reliant, Christian western democracy which is economically powerful, militarily intimidating, politically free, culturally vibrant and socially cohesive". Apart from political parties, there are citizen associations that support a strong Australia. An example is Advance Australia (https://advanceaustralia.org.au) whose aim is to "protect, defend and advance Australian values and champion institutions which strengthen the Australian way of life".

References

Bolt, A. 2021. Not one of the ABC's 4,600 staff warned Bruce Pascoe was a fraud. Herald Sun, June 14.

Jones, A. 2020. Australian defence bureaucrats "run for cover" amid Brereton Report. Sky News, November 9.

Pascoe, B. 2014. Dark Emu. Magabala Books, Broome, Western Australia.

Sutton, P. and Walshe, K. 2021. Farmers or hunter-gatherers? The Dark Emu debate. Melbourne University Press, Melbourne, Australia.

Index